Scientific publishing and presentation

Claus Ascheron

Scientific publishing and presentation

A practical guide with advice on doctoral studies and career planning

 Springer

Claus Ascheron
Heidelberg, Baden-Württemberg,
Germany

ISBN 978-3-662-66403-2   ISBN 978-3-662-66404-9   (eBook)
https://doi.org/10.1007/978-3-662-66404-9

This Springer imprint is published by the registered company Springer-Verlag GmbH, DE, part of Springer Nature.
The registered company address is: Heidelberger Platz 3, 14197 Berlin, Germany

Preface

This book teaches the skills of giving convincing scientific presentations and writing good scientific publications. These skills are important for presenting one's own results with success.

At colleges and universities, students are usually well prepared for the requirements of understanding science, conducting experiments and interpreting the results. Surprisingly, however, the scientific communication of the results obtained in lectures and journal articles is rarely the subject of courses at educational institutions. Most young scientists are simply thrown in at the deep end after graduation and exposed to later demands unprepared. They have to learn by trial and error during the new activity (learning by doing). A certain degree of perfection is only achieved through many mistakes and failed attempts. But as we all know from experience, the learning process necessary for this comes to a standstill for some young scientists at a very early and rudimentary stage. This can be seen in the often undeveloped ability to present and the poor quality of many publications.

After leaving the protective study environment, young scientists often have to struggle hard until they have mastered these initial difficulties and become accepted partners in the scientific community, not only by achieving excellent scientific results, but also by presenting them appropriately and impressively.

At many research institutions, it has been recognised over the last 20 years that it is also important for the education of undergraduate and postgraduate students to develop scientific communication skills. Soft skill courses are offered for this purpose. The author of this book gave such courses at numerous international universities and research institutions.

The aim of this book is to help young scientists at this early stage of their scientific career and to accelerate and facilitate this learning process by providing relevant advice and information in a concentrated form.

The book gives background knowledge on the process of scientific publishing and also shows the way how to write a dissertation in a structured way. Advice is given on planning an academic career afterwards.

This book is not a scientific textbook. It asks the often-forgotten questions about how to present your scientific results effectively and convincingly in presentations and publications.

If these advice are taken seriously, they will help to develop a good and efficient working style, which can be useful for a lifetime and will bring satisfaction in work.

Young scientists and advanced students are likely to derive the greatest benefit from this book, but any other scientist can also receive helpful suggestions for improving their oral and written communication skills and for better presenting their results and ideas. Even experienced scientists and professors with long scientific and teaching experience can benefit by critically rethinking the quality of their oral and written presentations in light of the recommendations provided.

The style of this book is casual and easy to read, avoiding confusing technical details. An extensive bibliography includes a number of further reading books on academic writing and presentation, in which the reader will find all the necessary facts, conventions and guidelines laid out in hair-trigger detail. Unlike some other literature that is overloaded with detail, the goal of this book is to present all the essentials in a concise, pleasantly readable and understandable manner without sacrificing the essentials. I hope that this goal is achieved.

The author gives the hints on the basis of rich experience in scientific work and communication, namely through

- Numerous publications in peer-reviewed international scientific journals
- Many presentations at international scientific conferences
- Teaching and lecture experience
- Helping colleagues and students prepare good publications and presentations
- Evaluation of a large number of manuscripts
- Critical attendance at countless conference presentations

The author has been invited by many universities and research institutes to give lectures and courses on the subjects of this book. Often at the end of them the comment was made, "It would be great to have this information in book form someday". Here they are. I hope you will enjoy reading it and that you can use this information to your benefit.

This book is the revised and updated version of the book *The Art of Scientific Presenting and Publishing,* published in 2007 [1].

Recognition

I would like to thank Dr. Angela Lahee for her collaboration on our English book *Make Your Mark in Science* [2], on which this translation is based, and Katharina Ascheron as "model presenter". I also thank my wife Regina Ascheron for her patience during the months I was busy writing this manuscript.

Heidelberg, Germany   Claus Ascheron
August 2018

Contents

1

Introduction

Becoming a well-known scientist and achieving major scientific breakthroughs is the secret dream of almost every young enthusiastic student. Some, but only a minority, will see this dream fulfilled. But not every young scientist is primarily motivated by the desire to achieve fame and glory – there are a number of equally significant motivations: one is scientific curiosity and the desire to understand nature; another is the goal of contributing to the great edifice of knowledge. Whatever your primary motivation, there is always one fundamental requirement: to become a successful scientist, you must possess the necessary qualities and skills.

1.1 How Do I Become a Successful Scientist?

The necessary personal qualities to be a good scientist include.

- scientific curiosity and the desire to explore and understand the world around us,
- the ability to think and evaluate logically,
- a certain mathematical inclination (especially in the natural sciences),
- the adequate memory of facts.

If you have these skills and have successfully completed the relevant training, then you have the potential to perform outstanding scientific work. But this

© The Author(s), under exclusive license to Springer-Verlag GmbH, DE, part of Springer Nature 2023
C. Ascheron, *Scientific publishing and presentation*,
https://doi.org/10.1007/978-3-662-66404-9_1

is only latent if it is not combined with a group of other skills. These skills, which can be learned in contrast to the above mentioned, are

- Presentation Skills,
- Writing and publishing skills.

This book is specifically dedicated to those skills. Reading this book will not teach you to become a better scientist, but optimizing your latent potential will make you a more successful scientist.

1.2   Presenting

After a scientist has come to some interesting new scientific results, the next challenge will be to communicate them in an appropriate way to the larger scientific community. If you keep your discoveries only to yourself, you will not contribute to the advancement of your field of knowledge, nor will you become known yourself.

The first step in disseminating your results will probably be to present them at an internal seminar or in a short talk at a conference. Some people, albeit the minority, have a natural gift for presenting. They can communicate well, present confidently and persuasively, and captivate an audience. Others are lucky enough to have supervisors who help them give better talks, sometimes skillfully copying their style so that it works from the start. But most scientists have to walk the rocky road of being subjected to repeated criticism from their peers and having to end their first talks with the feeling that most of the audience has completely lost interest in the talk. This learning through much and often failed trial and error works to some extent, but it can be a long and arduous process whose success is not guaranteed.

Good presentations are also enormously important in scientific teaching, in lectures. We have all struggled through bad lectures, which were incomprehensible in terms of content and/or acoustics, poorly didactic and supported with barely legible PowerPoint images *(slides)*, transparencies or blackboard images. So we left the lecture without having learned anything, and finally had to read the material afterwards in textbooks. This experience shows that not all older and experienced professors are good at presentation techniques either. After repeating the same mistakes year in and year out, they have become so ingrained that the person is no longer even aware of them. Therefore, it is a useful investment for a young scientist to try to perfect his or her presentation style early on. Chap. 3 of this book will help you to avoid the most common mistakes and to prepare and deliver convincing presentations.

1.3 Publishing

In practice, the main motivation for publishing is to document one's contribution to science and to obtain priority rights to discoveries. Scientists only become known and accepted in the scientific community through their publications. In English there is the expression "publish or perish". Publishing results also includes presenting them in conference presentations. If your publications and presentations stand out from the average, you will attract the attention of crucial scientific colleagues.

Therefore, writing good scientific publications and giving appropriate talks is an essential measure for young scientists who strive for recognition and establishment in the scientific community. Only some manage to enter the scene with a bang, such as Brian David Josephson or Rudolf Mößbauer, who received the Nobel Prize at quite a young age. For most scientists, the early days of a scientific career, if not the first decades, represent a patient process of gradual improvement in their status, often with unpleasant accompanying circumstances.

First, they must share credit for their discoveries with senior and superior colleagues. It may take them a long time to achieve the status of being the first or sole author of an article. But this is not necessarily a disadvantage. These early years are the best opportunity for a young scientist to learn from more experienced colleagues the tricks of publishing and publicizing their own results. But this process is usually not consistently pleasant.

As mentioned in the preface, the teaching of skills in communicating scientific achievements is not part of the educational program at most universities. Nevertheless, writing good publications and giving good talks are skills that can be learned. With the help of this book, you will quickly become familiar with the main criteria for writing good publications and learn to avoid the typical mistakes. Your colleagues will be surprised at how well your ability to write scientific articles has developed after you have received the chapter in question. Once you have made initial progress, you should take the time to compare your style again with that of your colleagues, from the point of view of the advice given here. You will see how much you have improved by taking into account some simple rules.

The chapter "Culture and Ethics of Scientific Publishing" (Chap. 4) has been included in the book to give the reader an idea of how modern publishing works. This additional information is useful for all readers who are involved with publications and publishers but are not sure what options are available to them.

1.4   Further Literature on the Topics of This Book

There are a number of other books on the above subjects, all of which have found their readers. Here I would like to refer to the following books, namely for

* Scientific Presenting on [3–9],
* Scientific publishing on [10–23],
* writing a dissertation on [24–33].

However, since they are not as detailed in content and give less detailed advice than the present book, it seemed reasonable to me to write this instructing guide for young scientists. The English-language book by Alley [9] comes closest to the intention of the present book.

2

Scientific Presentation

At a funeral, the average consumer would rather lie in the coffin than have to deliver the eulogy.
Jerry Seinfeld, *On the Fear of Public Speaking.*

Most students and scientists, whether they are experienced or just starting out in their careers, are required to give scientific presentations on a regular basis. Usually, more lectures or talks are given than publications are written.

As we all know, excellent scientists are often poor lecturers. Hermann von Helmholtz, for example, a brilliant scientist, was known for his poor lectures, which were only understood in the first rows of the lecture hall. Being a good scientist is no guarantee of having a natural talent for presenting well and at the beginning of their scientific career most researchers struggle hard with this task. A lucky few have a natural talent for presenting well; others learn quickly from their mistakes; but a great many scientists, unfortunately, continue to give bad talks throughout their careers. Stop.

The success of any presentation depends on the following main factors: (1) content, (2) visual aids, and (3) delivery style.. While the preparation of the content of a presentation requires the same knowledge and follows the same criteria as the preparation of a publication, the other factors – visual aids and delivery style – require completely different skills. In this chapter, recommendations are given on how to optimise the preparation and delivery of a presentation.

The aim is twofold: First, I want to make you aware of pitfalls and typical mistakes of which presenters are often unaware (whether students or

The original version of the chapter has been revised. A correction to this chapter can be found at https://doi.org/10.1007/978-3-662-66404-9_7

experienced scientists) but which are mercilessly criticized by the audience; and second, I want to give you the necessary information on how to plan, practice, and perfect presentations.

If you make your initial efforts to present your research findings in an appropriate manner, then you can significantly speed up the process of establishing yourself as a recognized researcher in your field. And you can also avoid criticism and wasted effort by giving good presentations. Your future students will be grateful to you, not only for presenting the information clearly in lectures and talks, but also for setting a good example.

Before going into detail, the different types of scientific lectures should be listed:

- oral conference presentations,
- Seminar presentations,
- Lectures for students,
- Defence of a qualification thesis,
- Defense of a project proposal,
- Poster Presentations.

All these types of presentations require the same skills. This is especially true for the visual presentation. However, they differ greatly in the amount of material selected and the level of detail in the discussion. Not only do lectures vary in length, from 10 min to over an hour, but the composition of the audience and their knowledge of the area being discussed varies greatly. These factors should be taken into account when planning a presentation.

In the following sections, unless otherwise noted, advice is essentially given on short talks and invited keynote talks at conferences, typically 10–30 min, and on defending a dissertation. For many young scholars and students, short talks are not only the most common type, but also the one they are most afraid of. Moreover, one should be aware that one will be critically evaluated by other external colleagues, which may include colleagues who will later have to help decide on a possible appointment. A good impression at the presentation can therefore have far-reaching positive effects.

2.1   Planning a Presentation

When you learn that you have a conference or seminar presentation to give, you will begin to think about what material should be presented. While it doesn't help much to tackle a talk months in advance, it is very useful to prepare it early enough, especially if you are relatively inexperienced. You will need a lot of preparation time, not just for gathering and selecting material, but also for planning and organizing the content of the talkand to prepare the visual presentation and practice the talk a few days before the big performance.

An acquaintance gave an excellent talk at a major international conference. When I congratulated him and said that his lecture was even better than that of the Nobel Prize winner who was present, he said: "You can't imagine how long I sat on the preparation of this lecture, the last two weeks I did nothing else".

You cannot say that he was quite inexperienced, because in the same year he received the Nobel Prize. But this shows that even the best scientists have to make significant efforts to prepare convincing lectures.

As a rule of thumb for preparation time, I recommend starting to put together a short presentation about 2 weeks before the conference. Much earlier is often not useful, since usually results of the last days before the conference are still included. For big talks, a little more preparation time should be allowed, at least 3 weeks. But all this depends on how prepared the material to be presented already is.

2.1.1 Audience

Before you start preparing your talk, you should ask yourself one basic question: Who will be the audience for my talk? What do these colleagues already know, and what are they likely to be interested in? Do you expect an audience qualified in your field or a mixed audience? For smaller specialized conferences, you can expect mostly specialists, while for large conferences with a broad range of topics, non-specialists may want to learn about other fields.

Or, for example, at a summer school, will the participants be primarily advanced students and young scientists? How far should the basics then be discussed and how far should the details be gone into? Most scientists forget to ask themselves these questions and are then surprised when the response to their after all so wonderfully prepared lecture is only moderate. The better you succeed in adapting the content to the expected audience, the greater the success of your presentation (Fig. 2.1).

When you speak in front of specialists more details can be presented. But for a more general audience especially the basics should be presented in a detailed and understandable way. This also includes not falling into technical jargon. The use of analogies to known experiences is always helpful and

Fig. 2.1 Respond to the interests and knowledge of the audience © Thomas Hauss, DGK

supports to achieve better understanding. In front of a general audience, do not go into too much detail and only present the important phenomena and broader implications in the most understandable way possible. Be aware that even the best prepared presentation will not be accepted if it is set outside the knowledge and interest of the participants. It is most complicated when you are speaking to an audience that consists of specialists and non-specialists specialists. In this case, it is advisable to structure the talk mostly as for non-specialists, but present more details towards the end for those who know the scene. A good mix is to use the first 50% of the presentation time for the general aspects and in the second half to present new things to the specialists. This way, every participant will leave the talk satisfied and feel that they have learned something.

You might also want to ask yourself what motivates the lecture participants to attend your lecture. Do they need the information you provide for their work? Is the content of the lecture for their studies or exams? Or do they just come to learn something new and enjoy a promising presentation? If you can answer these questions, it will help you not to disappoint the audience.

A lecture is most appreciated when the majority of the participants understand what you are presenting. Even if they already knew some of it beforehand, you should not forget the words of Enrico Fermi, who said: "You should never underestimate the pleasure people get from hearing something they already know".

A few years ago, based on an interesting submitted paper abstract, I was invited to give a keynote lecture at the Japan Society of Applied Physics conference. But unfortunately I could not get all the expected results by the time of the conference. So I decided the night before to give another talk under the same heading, exclusively on the new method developed by our institute. The lecture contained few new scientific results, but instead a very clear explanation of the experimental technique, its fundamentals and its potential for application. Surprisingly, the response to this lecture was much more enthusiastic than to my other, more specialized lectures. Five colleagues thanked me for the explanations given and said, "We have seen the director of your institute lecture on this subject several times before, but never understood what it was all for. Now, for the first time, it has dawned on us how significant this method is. We would like to develop a cooperation with you".

A fundamental mistake especially of many young scientists, is to assume out of respect that the other colleagues present are familiar with the research being presented. But this is true for only a few. Most researchers have little insight into more research areas than their own and a rather small number of familiar analytical techniques. Therefore, as the presenter, you almost always have the advantage of knowing more than your audience. Use this knowledge advantage to give the most interesting and instructive explanations possible, rather than trying to confuse and impress the audience with incomprehensible findings.

2.1.2 Title and Abstract

Normally you have to define the title and abstract of your presentation when you register your presentation, i.e. when the conference is announced and long before the presentation is scheduled. Since it is no longer possible to make changes after this, you should already think carefully at this early stage about what can be presented in the available presentation time and to what extent the expected audience is already familiar with the topic. This consideration is important for the title and the content of the abstract. If not all expected results are available in the early stage long before the conference, then the abstract must be should be formulated in a more general way. The same is true for colloquium presentations, which also normally require the prior submission of an abstract for the presentation announcement is required.

The title should be as short and appealing as possible. It should be clear from the title what the talk will be about. An attractive title is especially important for conferences with many parallel presentations. If an appealing title and a well-written abstract arouse interest in the conference topics then the presentation will be well attended. Conference attendees will only come to your talk if it promises to be interesting and informative. If you are working on a hot topic or/and have achieved impressive results, then don't be too modest in your choice of title and abstract wording.

Then, when you give the talk, you should also speak on the announced topic and not deviate too far from it. However, sometimes some modification is unavoidable, for example, if you could not get all the expected results in time for the conference. But one should not make a habit of raising expectations that the talk does not meet. It would certainly not be a good idea to announce a talk on X-ray methods and end up talking about nuclear physics.

When very prominent scientists are invited to give plenary lectures about their work, then they are usually free to vary the topic within certain limits. Many years ago, I met the Nobel Prize winner Gustav Hertz, who was already well over 80 years old, after one of his lectures (Fig. 2.2).

When I admiringly told him that I was impressed by how he still prepares and gives excellent lectures at such an advanced age, he replied: "Young colleague, for years I have been giving the same lecture over and over again. The colloquium or conference organizers just ask me to change the title a little so that it doesn't look like old hat, and occasionally to shift the focus a little."

Unfortunately, as a less prominent science star, you can't easily afford that luxury.

Fig. 2.2 Gustav Hertz © Claus Hampel/AP Images/picture alliance

2.1.3   Collection of Materials

If you should be speaking to a general audience, then you will need to elaborate a little further. In this case, you should focus on the basics and the more general meaning of the topic and present this as understandably as possible. If you are speaking to specialists, the general introduction should be much shorter. And you will present such content as you would publish in an article.

With this in mind, one should begin to draw up a list of content ideas. ideas. This does not require any particular structure. It's just a matter of first noting down everything that could be in the presentation. When you have written all this down, you should proceed to a division into two groups:

1. Content that needs to be presented to show the topic, and
2. Content that can be presented when time permits.

For this purpose, it is helpful to have all the diagrams to be used in perfect form. If you now think about the outline for the presentation, it becomes even clearer which information belongs in which group.

When preparing the content of the presentation it is recommended to have the following aspects in mind:

* What is the state of knowledge in your research area? Are you familiar with the current scientific literature?

This should be presented as an introduction and also used to explain the motivation of your work. If supporting or contradictory results have been published by other colleagues, you should at least know about them. Even if you do not mention these results in the discussion, you should at least be prepared for corresponding questions.

* What additional information should you provide in the presentation to ensure full understanding or to fully appreciate your efforts?

This could be something basic, something about the history, some interesting experimental details, or background information. Interesting side information generally increase the attention for your presentation.

* What are the main information? Does the time available allow all of it to be displayed?

If this is not the case, then the list of necessary and possible contents should be critically reviewed again.

- Is there enough significant new information in the presentation?

The audience will leave the presentation disappointed if you have not presented anything new at all. A presentation must always offer something new compared to what has already been published or presented at other conferences. Be aware that in a short presentation or an invited keynote presentation, your own results must predominate in terms of content.

The situation is somewhat different for overview lectures, which are often plenary lectures. are. They present an overall assessment of what other colleagues have produced in terms of relevant results. The speaker's own contribution may be of secondary importance. However, at least an attempt should be made to introduce interesting new points of view. Once you have considered all this, the next step should be to develop the detailed plan for the presentation in writing. The sequence must be both logical and didactic, so that the audience can follow the presentation without effort.

2.1.4   Structure of a Presentation

The structure for a short talk could be, for example:

- Introduction, state of knowledge,
- the unsolved problem, motivation,
- Experiments to clarify the open questions,
- Results
- Interpretation,
- open questions for future research,
- Conclusions/Summary.

As a structure of a longer invited overview or plenary talk would be possible:

- What is the XYZ effect?
- Why is it important and interesting?
- What open issues are currently being worked on?
- What is my contribution to this mosaic?
- Summary of the main findings
- What picture emerges from this?
- foresight

It is not the purpose of such a structureto determine the final content, but merely to provide a guide to help you organize the material and your thoughts into a coherent story that has a beginning, a middle, and an end. You should try to tell every talk, long or short, like a story. We know that some stories end happily and others leave the reader or listener with unresolved questions. Therefore, don't feel bad if you can't offer a definitive solution to all the unanswered questions.

If the title did not have to be decided long beforehand, which is usually done when registering a presentation, often before all the expected results are available, then you can now still think of a good presentation title which in turn can help to set the focus correctly and to communicate the information as comprehensibly as possible.

After all these preparatory steps, we now come to the core.

2.1.5   Detailed Structure and Content

The rough structure presented above of the presentation for the selection of material is not yet sufficient as a fine structure for your presentation. For a conference presentation the structure should be the same as for a scientific publication, whereby logic and didactics of the presentation are important. A good outline and structuring of your presentation is also understood as an expression of your clear thinking.

However, if a lecture were to be given in such a condensed form, as if a publication were being then it would be very difficult for the audience to follow. In an oral presentation, the information must not be presented in such a densely packed way. When you read a publication, you can think about it step by step, read a sentence a second time, go back in the text, reread elsewhere to clarify terms or basics. This is not possible in a lecture. Here, everything must be immediately understandable. Therefore, a lecture must contain many more words than one would use to present the same results in a publication. Such a detailed explanation is only possible in the given time if one restricts oneself to the essentials and does not overload the lecture with content. A typical mistake, especially of young scientists, is to want to present everything one knows and has researched, which does not leave enough time to discuss everything touched upon in sufficient detail. Therefore, the advice is:

For a scientific lecture, less is more.

The didactic element. is even more important for a lecture than for a publication, even if you are lecturing to specialists. For this it is important to use analogies, to choose a simplified and straightforward line of argument, and to visualize the information as much as possible. If you are speaking to an audience that is unfamiliar with specialist scientific terminology make sure you don't fall into jargon that will put the audience off. Try to present the content as simply and as interestingly as possible. It should be your goal that the audience leaves the lecture with the feeling that they have understood everything and learned some new things.

The next step must be to logically divide the material and distribute it on PowerPoint images *(slides)*, to break it down into digestible individual morsels.

The goal here is to divide the material well, with each slide should represent a part of the whole story as coherently as possible. You must be aware that you will need approx. 1–1.5 min for each individual image to provide a comprehensible explanation. Give the audience the necessary time to absorb the information. Keep in mind that you need to explain the figure contents, axes and trends. This means that for a typical short talk at a conference of 15–20 min duration, you should plan no more than 10 to 15 content slides plus a title slide and an acknowledgement slide.

The order of the slides is determined by the structure that has been considered beforehand. Some of the main points of the presentation will require more than one slide. In the section on slide preparation (Sect. 2.2.2) you will see in detail how much information should be shown per slide. Subsequent sections will go into more detail about slide preparation.

It is not advisable to write down the lecture text word for word. The pictures will help you remember the main points. But you should think carefully beforehand what you want to tell, plan the speech in detail. Beginners need to make more notes about what to tell to the pictures than experienced speakers.

The structure of a lecture is similar to the structure of a publication. It should contain the following parts:

Introduction – What Is the Problem?

A frequently heard introduction is: "Today I would like to tell you about our work on …", where the title of the talk is spoken again. This is a safe but rather undramatic way to begin a talk. For long talks, it is often useful to preface with an overview of the content. For short talks, however, this is a superfluous luxury that costs some of the scarce talk time. If you want to make it clear at which point in the outline of an extensive lecture covering many different problems, you can also briefly display the table of contents for each new topic, with the current topic being with the current topic highlighted in color. For this, however, the table of contents should be displayed in only one slide when concentrating on the main headings should be presented in only one slide. After the introduction, the topic is briefly outlined, with the motivation for the work carried out is given, perhaps based on the limited knowledge achieved so far. The questions to be answered by the presentation should be stated.

Have you ever noticed that it is often the first three sentences of a speaker that determine your further interest? This often influences whether you are prepared to pay serious attention to the lecture. to the lecture. Therefore, you should strive to make the opening sentences as interesting as possible. If you are ever distracted by a dull or confused introduction or by speaking monotonously or too softly, it will be difficult to regain your attention. Therefore, you should consider the introductory sentences very carefully and perhaps practice their perfect delivery beforehand. A funny reference to obscure results in an initial slide or a humorous remark can also help break the ice. But if you're not familiar with humorous interludes… then you are on the safer side with a clear presentation. Through your introduction, each participant must come to believe that you have important things to share and are highly competent – not to mention that you are convinced of what you are telling.

Experiment and Methods – What Do We Do?

In this section, which is usually brief, you will describe the methods and analytical techniques you have used. If they are well-known methods, it is sufficient to mention them; for self-developed, less well-known methods, explain a little more. Also, the sample preparation and processing of the primary data should be made clear so that the results obtained do not appear mysterious (Fig. 2.3).

Results and Discussion

Sync and corrections by n17t01 like in a publication, this part is the most important. In scientific papers, it is sometimes divided into two separate sections. In a lecture, however, the results must be discussed when they are presented and while the impression of them is still fresh. Another difference between an original scientific paper and a short talk is that in a publication,

Fig. 2.3 Experimenter Marie Curie © The Print Collector/Heritage Images/picture alliance

extensive data can be presented with a long interpretation. In a short talk, there is not as much time available for this. Moreover, this could overload the audience with information. Therefore, it is extremely important to select the most significant results and convey them clearly and convincingly. When you show your results images, take the time to show the axes and the curve. and the curve and everything you see in the picture, also explain it in words – and not too quickly. Don't hang your audience out to dry by presenting the results too quickly. Give him enough time to let your explanations sink into his consciousness. The goal should be that the audience has to think as little as possible and understand as much as possible.

When you present your results, you should discuss them in the light of the results published by others, in order to place your new knowledge in the existing body of knowledge, so to speak. existing body of knowledge. However, you should not – as is usually the case – only use findings that support your own interpretation. interpretation. If different conclusions have been reached in comparable studies by colleagues, you should at least mention this and at the same time try to explain the reasons for the existing differences. But never give the impression of presenting nothing new and perhaps only discussing literature results. Always make it clear what is essential and new in your work.

Always try, as far as possible, to provide a clear interpretation of your results. But if the results at hand are not yet sufficient to give a conclusive, final interpretation, then you had better list the questions that are still open than to fall into wild and unsubstantiated speculation. Stating a reasoned working hypothesis in conjunction with a plan to confirm it might be a more useful lecture ending in such a case than indulging in wild guesses.

Conclusions – Summary and Take-Home Message

At many lectures you experience a summary at the end. This systematically repeats the main points of the discussion without saying anything new. Often this means repeating verbatim the main propositions of the discussion. An alternative to this, and the better way, is to conclude by drawing some generalized conclusions.

In doing so, try to evaluate your results from the distant view, so to speak. while in the discussion you had the close-up view. in the discussion. In doing so, highlight the generally significant aspects and implications of your work. Try to convey a message to the audience that they can take home with them. Phrase this message as simply and concisely as possible and in short sentences. Since most people can't remember more than three things at once, limit your core message to just three points if possible. Then your presentation may be the only one of the conference session that participants remember afterwards; and there will be a feeling that your presentation was one of the highlights of the conference.

Recognition – Who Contributed to or Funded the Work?

It is good practice to mention with thanks at the conclusion of a presentation those colleagues or organizations that contributed to obtaining the results presented without being identified as co-authors. This may include staff, technical assistants, students, discussants and the boss who created the conducive conditions or at least allowed the work to be carried out, even if he did not directly contribute to it himself. Mention should not be made of the institutions that financially supported the project.without which the research budget would not have been available. Since at some conferences invited keynote papers may have only one author, it is then essential to acknowledge the support of colleagues who would otherwise have emerged as co-authors. In such cases, it seems very fair if the presenter begins his or her remarks with a photograph of the group that produced the results to be presented, representing the individual contributions of colleagues or students. The advantage of a prefatory acknowledgement is that the audience is not distracted from your main message in the conclusions. These can then remain on display throughout the discussion and thus better sink into the minds of the participants. Occasionally, the acknowledgements are also shown is shown immediately before the conclusions and remains projected on the wall during the discussion.

The acknowledgement is usually the last slide of a presentation. This text does not necessarily have to be read out, unless you want to give special recognition to certain contributions.

However, if you have worked completely independently and without interaction with other colleagues and have not had any external sponsorsthen, of course, the recognition at the end of the presentation is not applicable.

Psychologically Seductive Lecture Structure

There however, there is also a completely different approach to lecture structure which, although I do not recommend it, is practiced by some scholars, often older colleagues. This lecture structure is as follows:

- In the first part of the lecture, you tell the audience what they already know. In this way, each participant gains the satisfying feeling of having some scientific understanding after all.
- In the second part something new is told. So everyone sees that attending this lecture is not a waste of time and you can learn something from it.
- In the third part, everything is presented in the most complicated way possible, with the aim that nobody understands anything at all. In this way, it is made clear at what high scientific level the lecturer is working, and respect for the lecturer's high science is maintained.

2.2 Visual Aids

2.2.1 Advantages of Visualised Representations

A scientific presentation is not something like a politician's speech, a pastor's sermon, or a radio broadcast. For a scientific presentation, the visualization of what is being presented is eminently important. The visual style of a presentation can contribute significantly to its success or failure.

There is a saying that goes, "A picture is worth a thousand words." This is especially true when it comes to presenting complex scientific and technical content. The pictorial representation of your results is much more efficient than in long tables and columns of numbers or only describing them verbally. Trends and dependencies become much more recognizable in graphical form.

The recommendation to use imagery also applies quite analogously to speaking. Where appropriate, try to speak in images and analogies. analogies. When existing knowledge and experience can be linked to, it is much easier to understand something new. Some people learn better by reading, others by listening.. If your information is presented both orally in the lecture and in writing in the projections, then both the sense of seeing and the sense of hearing are addressed. as well as the sense of hearing. By addressing these two sensory channels simultaneously, your message will be received more intensely. Even the sense of touch can be additionally involved, for example by passing samples around the auditorium. Furthermore, the senses can be addressed by simple and impressive demonstration experiments. can be addressed. A notable example of this is Richard Feynman's demonstration of the brittleness of a rubber sealing ring at low temperatures to explain the Challenger disaster.

Demonstration experiments are also very important for teaching purposes in experimentally oriented lectures. are very important. As Jearl Walker of Cleveland State University said, "The way to get students' attention is to demonstrate experiments where there is a real possibility that the professor will have an accident".

In most lectures, however, only the visual information projected on the wall behind the speaker. Computer presentations are often much more suitable than writing on a blackboard with chalk.

Computer Presentations

Here screens are projected onto the wall with a beamer. projected onto the wall. Typically, Microsoft PowerPoint a component of the Office package. This is also available for Macintosh computers. But Mac fans will probably prefer the Apple presentation program Keynote presentation program. Everything said about PowerPoint in the following text applies just as much to Keynote and comparable presentation programs.

Computer presentations have a number of advantages over the old slides. These programs are quite simple and intuitive to use. If you use a predefined standard layout or a self-generated layout, then all slides automatically have the same format, e.g. with regard to font and heading size and colour, background colour, built-in institute logo or name of the presenter. During the presentation, you then only have to press one button to move to the next image.

The didactic advantage of such a presentation is that images can be shown step by step without overwhelming the listener with too much information that only becomes relevant later. This is useful, for example, when complex images are built up or several curves are presented in one figure. If you fade in each element or curve in turn and give the explanation of the differences throughout, then the relationships can be brought out much more clearly. Similarly, blocks of text or lines can be faded in only when they are discussed. This prevents what is to be discussed later from being read beforehand, thus distracting the audience's attention from the spoken word. However, a line-by-line structure of the slides, the fluidity of the presentation may suffer because the presenter is not sure what the next topic to be discussed will be.

The use of colours there are technically no limits. But one should be aware that too colorful images tend to distract rather than enhance attention. The color should support and emphasize thestructure of the presentation. It is useful and loosening to use different colours for the different headline levels. different colours for different levels of headings. However, the choice of colour, font size and type should then be kept consistent throughout the lecture. Also, the main points can be highlighted in color. However, you should definitely make sure that the colored texts and curves are easily recognizable. Therefore, only intense colours should beused and not weak colours such as yellow or light blue. The slides have landscape format. Animations, moving pictures or cartoons can be used to liven up a lecture.animations, moving pictures or cartoonscan be used. But you should check the use of such animations very critically. Often they are rather disturbing and distracting.

Before you give a lecture, you should make sure that there is a suitable projector on site. In the case of computer presentations, it is highly advisable to have checked for some time before the presentation whether it will be possible to connect your own computer to the projector. In many conference presentations, the speaker needs a considerable amount of time to solve this connection problem. ...to solve this connection problem. Therefore, I recommend: Have your presentation on a memory stick... with you. If you have connection problems, you can borrow a colleague's computer where it worked before and give your talk without losing too much time.

Another advantage of computer presentations is that you can make last-minute changes and additions. at the last minute. Imagine that a colleague is presenting something in front of you that relates to your following presentation. If you have your notebook with you, you can add additional points to the presentation. Then everyone will be impressed by how up-to-date your presentation is, that you are already discussing results that were presented only 10 min ago.

2.2.2 Preparation of the Slides

Poor preparation of the visual material of a presentation is often the cause of audience discomfort. The following guidelines should help you to optimise the visual aids and to achieve a pleasant visual impression of your presentation. Please note that it is only sensible to prepare the slides until the content of the presentation is clear – if not on paper, then at least in your head.

2.2.3 How Many Pictures?

The allowed presentation time determines the possible number of slides. As already mentioned above, as a rule of thumb you should assume approx. 1.5 min per slide. per slide. You don't just show the picture, you also have to tell something about it, except self-explanatory cartoons to loosen up. Give the audience enough time to understand and absorb what is being shown in terms of text and images. Under no circumstances should you go below 1 min per slide. If you know in advance that there is a lot to say about some of the images, then you should also allow more time for this and take this into account with regard to the number of slides. take this into account. Accordingly, for a 15-min presentation, you should plan no more than ten slides plus a title and acknowledgement slide. should be planned. Once the content has been determined, divide it into units so that each slide covers a related topic. There should not be more than one topic per slide. It is better to have more slides than to cram too much information into one slide. Be aware that you will not take more time in the presentation if you spread the densely packed information of overloaded slides over a larger number of slides that are less densely packed in terms of content. When planning, of course, do not forget to prepare an introductory slide and a summary slide.. Do not start in medias res, and do not leave the end of your presentation open.

2.2.4   Amount of Information

Probably the most important advice to give here is this: don't overload the slides. The smaller the amount of information per slide, the better the information will be absorbed. Imagine that a presenter wants to convince his audience of the falsity of a hypothesis. More impressive than a slide with ten complicated and incomprehensible formulas is a slide with only the word "No". While this is shown, a few words can outline the reasoning better than rampant detail.. This is an extreme example, of course, and in many cases more explanation is needed. The message "less is more" should always be kept in mind when preparing the material.

A golden rule established by learning psychologists says that a slide should contain no more than 13 lines of text if the audience's attention if the audience's attention is not to be overwhelmed. Some presenters prefer only five lines to keep the units of meaning manageable. manageable. American psychologists go further: they say a slide should contain no more than 30 words. These are easily digestible smaller chunks of information that can be on only three to five lines, but this means you have to break the information down into very small chunks.

If you plan to summarize the information in paragraphs then about five paragraphs per slide is a golden measure.

Only show text and images that you are actually discussing. Therefore, remove parts of the text and images that are not discussed, as this would only create discomfort among the audience.

2.2.5   Headings

The best structure is to have one heading per slide. If there are several slides on one main topic, there can also be a summary heading above it. The headlines should be short and concise and should also contain attractive keywords to keep the to arouse the interest of the audience. A short question also makes a good headline. For example, after clearly stating your new hypothesis, the next slide could be titled, "How can we prove this?" – and then explain the experiments you performed. However, like all good ideas, this one should not be overused. If you are unsure, then a less attractive but boring correct title is still better than an attractive but misleading one.

2.2.6 Key Words and Key Findings

The main text content of the slides should consist of bullet points. consist of bullet points. The fully formulated text has to come from the presenter. Under no circumstances should you formulate complete sentences. These have already been read by the audience long before the speaker reads them out, which actually makes the reading out becomes superfluous. As a result, even the reader can be perceived as superfluous.

On the other hand, the keywords should already contain a clear message and should not appear dubious. Better than just writing "cell division", for example, is to include some additional, relevant information, such as "reduced cell division after two days".

The texts of the slides do not have the sole purpose of telling the audience the main facts. They are just as important for you as the presenter not to forget everything you wanted to discuss. If you have prepared your slides thoroughly, then you don't have to worry about forgetting to discuss essentials.

2.2.7 Typography and Font Size

Use different font sizes and fonts (fonts) to distinguish the different hierarchical levels of headings and the rest of the text. If you introduce a third font size/font in addition to two levels of headings and the main text, note that there should not then be any more, as this could introduce too much clutter into the picture (Fig. 2.4). Slant (italic) should be treated as a separate font. Also, keep the same font and heading scheme for the entire paper. Always adopt the

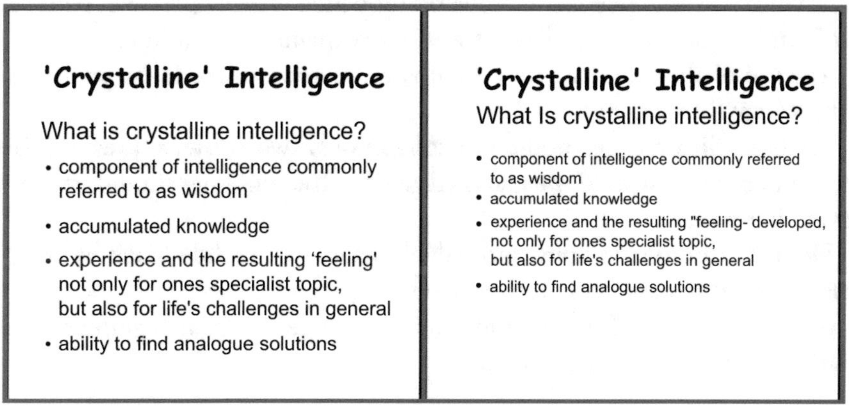

'Crystalline' Intelligence

What is crystalline intelligence?

- component of intelligence commonly referred to as wisdom

- accumulated knowledge

- experience and the resulting 'feeling' not only for ones specialist topic, but also for life's challenges in general

- ability to find analogue solutions

'Crystalline' Intelligence

What Is crystalline intelligence?

- component of intelligence commonly referred to as wisdom
- accumulated knowledge
- experience and the resulting "feeling- developed, not only for ones specialist topic, but also for life's challenges in general
- ability to find analogue solutions

Fig. 2.4 Appropriate and too small font size in a slide

same font, size and colour the same level of hierarchy. Likewise, the numbering should be consistent throughout the presentation.

What font size should be used? The size of the lecture hall has an influence on this. To ensure that your slides are easily visible from all seats, the minimum font size should be 18 point. 20 point is even better.

Different fonts and colors can also be selected to set focal points. set. While slanting is the first choice for emphasis in an article, boldface or highlighting in red is best for a lecture. or highlighting in red. Why this difference? A printed text is usually read in its entirety, with slanted highlighting standing out. Here, bold would distract the eye from the rest of the text. With a slide, however, that's exactly what you want to achieve, namely to focus on the most important statement. Therefore, bold print or coloured highlighting is the most is most suitable.

2.2.8 Diagrams

Diagrams and other images must also be recognizable and legible. This means that all lines must be strong enough and intense enough in color to be clearly seen. Color should be used to distinguish or highlight individual curves or parts in images.

Just like the text, the images should also should be limited to the main information. Do not show all 36 steps of the semiconductor cleaning procedure in one picture if this is only marginal information for the lecture. While this may be useful additional information in a publication, it will only distract from the main point in the lecture. Select only relevant information.

In graphical representations, all axes must be clearly named, also with at least 18 point font size. If this is not the case with scanned images, relabel the axes, which is easily possible in PowerPoint which is easily possible. Non-self-explanatory photos should have at least one explanatory text line.

The use of the sans serif font Arial is more appropriate in diagrams than Times New Roman.

Scientific illustrations should not contain only two arrows as axes. It is better to put a frame around the graph (Fig. 2.5). The unit lines should be drawn on all four sides.

Do not combine too many individual images in one slide. Each individual image should remain easily recognizable.

Sometimes it is useful to include an inset smaller image in a graph to schematically illustrate the measurement.

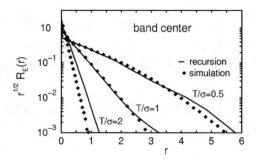

Fig. 2.5   Figure with frame

If possible, try to show a summary key illustration towards the end of the talk.

Longer texts are easier to read on a light background than on a dark background.

2.2.9   Mathematical Formulae

In a lecture, only such material should be shown as is either known or easily understood by the explanations given in the lecture. Therefore complicated mathematical derivationswhich require a lot of time to comprehend, should better not be presented in lectures (Fig. 2.6).

For this purpose, a publication is more suitable. If a mathematical derivation is the main component of the work, then it is better to show the result and outline the way in words.

There are, of course, always scientists who cannot resist the temptation to show off their superior mathematical education. They mistakenly believe that colleagues and professors can be impressed by a complicated presentation of the simple. They think that the elegance of formulation.

$$\ln(e) + \sin^2 x + \cos^2 x = \sum_{n=0}^{\inf} 2^{-n}$$

is in no way achieved by the equivalent simpler formulation $1 + 1 = 2$.

Your audience will appreciate it more if you follow the KISS principle:

Keep It Short and Simple ("keep it short and simple").

$$'q'_{13} : T_{(y_o, p_x, p_z^a)}(Y \times T^*(X \times Z)) \hookrightarrow T_{(p_x, y_o, p_z^a)} T^*(X \times Y \times Z)$$

is transversal to $\mathrm{Im}(\delta_\pi^{-1}(\lambda_1 \times \lambda_2) \xrightarrow{\ 's\ } T_{(p_x, y_o, p_z^a)} T^*(X \times Y \times Z))$, where $y_o = \pi(p_Y) \in T_Y^* Y$. Now, let $\rho' \subset \rho \subset T_{(p_Y^a, p_Y)}(T^*(Y \times Y))$ be the isotropic subspaces defined by

$$\rho'^\perp = T_{(p_Y^a, p_Y)}(Y \times_{Y \times Y} T^*(Y \times Y)) \text{ and } \rho = T_{(p_Y^a, p_Y)}(T_Y^*(Y \times Y)) .$$

Then we have:

$$T_{(p_x, p_Y^a, p_Y, p_z^a)}\left(X \times Y \times Z \underset{X \times Y \times Y \times Z}{\times} T^*(X \times Y \times Y \times Z) \right) \simeq E_Y \oplus \rho'^\perp \oplus E_Z^a ,$$

$$T_{(p_x, y_o, p_z^a)} T^*(X \times Y \times Z) \simeq (E_X \oplus E_Y^a \oplus E_Y \oplus E_Z^a)^{(0 \oplus \rho' \oplus 0)} ,$$

and,

$$T_{(y_o, p_x, p_z^a)}(Y \times T^*(X \times Z)) \simeq (E_X \oplus \rho \oplus E_Z^a)^{(0 \oplus \rho' \oplus 0)} .$$

Fig. 2.6 Mathematical overkill

Are perfect slides always the most appropriate? In a mathematics lecture, complicated and long derivations and formulas are better understood if they are developed step by step with chalk on the blackboard than if they are perfectly projected on the wall all at once as a *fait accompli*. This bludgeons the students.

2.2.10 Use of Colour

Your slides should be pleasing to the eye and direct the viewer to the essentials in a natural way. Therefore, color should definitely be used, but sparingly and supportively and not as a purely illustrative end in itself. A text presented in a rainbow-like manner contains no more information than a black text. Too much color distracts rather than helps focus. The colour must serve a purpose and highlight headline structure, information, or essentials. However, unusual color combinations such as red text on a green background should be avoided. Not only would this challenge the eye and emotion of most listeners, but it would also discriminate against a few listeners. This is because about 10% of all men (much less common among women) are colour blind.

The same applies to the use of colour in pictures. Here, many lectures make the mistake of using colors that are too weak and hardly recognizable. It is best to use intense colours such as black, red, blue, green, brown. If you have to use many different colours, e.g. to illustrate any intensity gradients, do not forget to show an explanatory colour legend. For delineations in schematic pictures, the adjacent use of contrasting colors is recommended.

2.2.11   Disclose Information Step by Step

Some presenters have a bad habit of filling their slides with too much text. It is therefore better to limit the information content to smaller units that can be shown at once.

With PowerPoint presentations the current information is contained in a slide which keeps the concentration on what is being shown. This is also particularly useful for the step-by-step didactic build-up complicated images. But the presenter may have a problem with the line-by-line presentation: Do you always know what's coming next? This can make such a lecture seem less fluid than one where more coherent information is shown in one slide. If you are in doubt yourself, it is better to show complete slides with not too much information.

2.2.12   Animations

PowerPoint offers a variety of possibilities to structure the text, to choose image transitions and to show moving images. The text can be built up letter by letter or line by line. The next image can be built up in a chessboard-like manner, from the right or left, from the centre or from the fog. Pictures and diagrams can be treated in the same way. As has already been stated in regard to the use of color, it should be used very sparingly; there must always be a supporting purpose in the foreground. If such animations are overdone, they tend to distract attention.

For example, my daughter once had to give a seminar presentation. She had prepared it using all the possibilities of Power-Point. of Power-Point: Pictures colored like Picasso, all kinds of picture transitions, line by line text structure, important messages brought letter by letter with a hammer sound ("and I want to hammer that into your brain"). Since the content was perfect too, she thought she'd get an A for it. But her professor's comment was, "This was an extremely impressive lecture. However, I cannot give you a grade on it. Because my entire attention was taken up by following your visual presentation, so I couldn't concentrate on the content at all."

The moral of this story is: you shouldn't exhaust all possibilities. A sparing, didactic use of color… and not too many animations in the transition from slide to slide support the presentation better. The presentation must underline the content and not take on a life of its own.

Animations should only be used sparingly to support the content without becoming an entertainment element in its own right. Sometimes, however, it has a loosening effect when a cartoon is shown. There is a wide range of cartoons available on the Internet. But even here it is important that the loosened up presentation has to support the message in a certain way.

2.2.13   Checklist

Before you leave your office for the conference, you should check that you have all the following with you:

* Computer with data and programs,
* all necessary connection cables and adapters,
* Memory stick with the lecture – in case the computer breaks down or cannot be connected to the beamer.

2.3   Practice Before the Big Event

If you have taken everything that has been said so far seriously, your presentation should now be perfectly prepared. The appropriate amount of information has been selected and the visual aids have been prepared in an appealing way. But this is only the first step. Good content and perfect visuals alone are no guarantee of a successful lecture. Equally important is its good oral presentation. If one is not very experienced in giving speeches, it is useful to allow a few days of practice before the big event. A few people have an innate talent for presenting and have no trouble captivating audiences and delivering lively and persuasive speeches. On the other hand, there are also many colleagues, including experienced professors, who have no presentation talent at all and do not make the slightest effort to improve their presentation skills.

In the lectures you attend, don't just look at the content, but look deeper, explore the structure of the lecture and reflect on the style of delivery. It is easy to learn from negative examples. It is more difficult to draw conclusions for your own presentations from excellent other presentations. Most of the time you don't think about why a talk was so excellent. You will find that you become a better presenter by becoming a more attentive listener. In this way, you may also be able to take a step in the direction of becoming a top scholar.

Therefore, anyone who is not convinced that he or she can give a good talk should practice before the performance. Such a "dry run"… I recommend at least for your first speeches. Another advantage of excellent preparation and extensive training is that you will be less nervous when you give your talk, because you know you can do it. because you know you can do it. If you have prepared your slides well, you will have bullet points for everything you want to say. for everything you want to say. You then don't need to fear leaving out anything important in the presentation. Of course, there is also the option of having an extra cheat sheet in hand. But this can rather disturb you in your presentation concentration.

Here are some more tips for reducing restlessness., which arises for the following reasons:

2.3.1 Are You Afraid of Forgetting Essentials?

In order not to get nervous, you should be aware that you are the only one who knows what you have prepared. Even if you leave out some of the prepared material, it won't bother you much if the presentation is still self-contained. It is better to leave out 50% of the prepared material and reach 100% of the audience than to present everything and reach no one.

2.3.2 Does It Worry You to Possibly Exceed the Lecture Time?

You can easily check the expected presentation time easily check by giving the talk to yourself in the same way as at the conference. In this process, the content can then either be reduced or extended even further. Also, the additional time for short interposed questions and possibly speaking a little more slowly should be taken into account.

If you watch the faces of the participants during your presentation, you will see whether your speaking speed is correct, or whether you should leave a little more time to digest the new information.

There can always be unexpected delays. Therefore, it is highly recommended to have more than one option to finish the talk earlier than originally planned. Structure your presentation so that the main information is contained in the first 75%. If need be, you must be able to leave out the next 25% of the planned content, which is supplementary to the topic. Or just have two to three slides towards the end that you can leave out without significant loss of information. However, it is somewhat distracting in a presentation if the slides are just clicked through. With such a preparation it depends then only on the clock what everything is presented. What you should never leave out are the conclusions at the end of the presentation. at the end of the presentation. Give the audience a powerful take-home message. Also, do not be pressed for time in delivering this most essential message of your talk.

2.3.3   Are You Worried About Not Finding the Right Words?

This can really become a problem if you are inexperienced in presenting or have to give the talk in a foreign language that you don't feel completely at home in. One way of crisis management could be to write down the talk and learn it by heart. If you have learned the lecture well – and are not interrupted – then your audience will not notice that you are only lecturing from memory and not freely. But preparing this version of the lecture takes a lot of time, at least a week of memorization and practice. But actually, this way of giving a lecture is not recommended. A learned speech is not usually as convincing usually not as convincing as a free speech. as a free speech. Also, an unexpected interruption or interposed question can make one can make you lose the thread completely. The better prepared your talk is and the more often you have given talks, the more confident you will be in finding the appropriate words. If you can't think of the foreign word you want to use, then simply try to paraphrase the subject matter in other words.

But you should never make the cardinal mistake of clumsy scientific lectures: reading a lecture off the manuscript. This is the worst way to give a lecture. A scientific lecture must always be a free lecture! a free lecture! Even a free lecture with some forgivable mistakes is better than a read-out lecture, where the reader's nose is constantly buried in his papers, the whole thing is delivered as monotonously as possible and there is no eye contact with the audience at all.

With a head of state who can provoke diplomatic conflicts by a single wrong word, the verbatim reading of a speech is necessary. Scientists, fortunately, are more tolerant of not quite correct choice of words.

2.3.4   Should You Feel Too Safe?

If you're going to give the same speech over and over, don't succumb to a false sense of security... succumb to a false sense of security. It is always useful to look at the material again before the presentation and to think about what you might need to add and at what point. The same presentation to different audiences requires specific introductions to keep them interested. Also, a different presentation time should be taken into account.

2.3.5 Do You Have Stage Fright?

Make Realize that no one knows about your topic as well as you do and you can't be cornered on the core content of your talk. Being well rested a relaxing walk before your presentation and taking a few deep breaths to relax immediately beforehand will help to reduce nervousness. nervousness.

2.4 Giving a Lecture

When you give a presentationyou should be aware that you are not only presenting content, but also yourself. Most presenters are not aware of this essential aspect at all. The audience will not only form an opinion about your work, but also about you as a person, your skills and competence, your scientific enthusiasm and your trustworthiness. …and trustworthiness. So perform as well as possible.

Participants will appreciate your presentation if it is rich in content, informative, clear, interesting and – especially in longer presentations – entertaining as well. Your audience gives you a valuable asset: their time. The participants of the lecture do so in anticipation of being compensated with interesting and useful information. They are also ready to be convinced by your story. But before you can convince others, you must first be convinced yourself. You must be sure before the lecture that your experiments., calculations and conclusions are correct. When you present something that you yourself are convinced of, the presentation is automatically made with more confidence, seriousness and certainty.

But seriousness alone is not enough. The following sections explain other key factors that go into successfully delivering a speech or lecture. You should keep these aspects in mind when practicing your presentation.

2.4.1   Manner of Speaking and Style of Presentation

Free Speech

As already discussed above, it is eminently important that you give the talk freely and not read from a manuscript. If you read from a manuscript, then you will inevitably not have eye contact with the audience because of the bent head. You could just as well speak to the wall. A read-out speech is usually delivered in a more monotone voice than the free speech, so the audience's attention suffers. If you have the right cues in your slides and know what to report on them, then there is no need to read off a speech.

But giving a free lecture in a lecture hall is not the same as telling someone something in a private conversation. It requires a lot more attention to clarity, speaking rate and intonation. Your voice should be understandable even in the last rows. Therefore, you need to speak loudly enough, with good emphasis, and clearly. If you are speaking in a large auditorium without a microphone. you should ask at the beginning whether you can be understood everywhere and, if necessary, speak louder afterwards.

Rhythm and Dramaturgy

The Speech rate should be slower than in normal conversation. The slower speaking is necessary on the one hand to avoid sound overlap in larger halls and on the other hand to give the audience enough time to understand and absorb difficult and new information. There is a danger that participants who have been left behind in their understanding will be lost for the rest of the talk. Therefore, short pauses at the end of significant statements are as important as speaking slowly. This should be done not just for dramatic effect, but mainly to help the preceding message be absorbed. It can also be helpful to repeat an important statement in different words and from a slightly different angle (Fig. 2.7).

Also remember to give your voice a certain modulation. that is, to emphasize the important things accordingly. In this context, it is useful to look at the habits of successful speakers. You will notice that they speak with a well-balanced dynamic. vary the rate of speech and use the intensity of emphasis skilfully. skillfully. Don't underestimate the importance of these aspects in how your message comes across to the listener.

As a scientist, you don't get the training in rhetoric and dramaturgy as an actor, priest, Roman senator or politician. This is reflected in scientific

Fig. 2.7 Presenting convincingly

lectures. However, scientists, almost all of whom have underdeveloped rhetorical and theatrical skills, are very tolerant of less than perfect lecture styles. Therefore, you have nothing at all to fear if you don't put on a shining act. With this knowledge in mind, and after hard practice... you will soon gain the confidence and poise necessary to deliver a convincing speech.

Language and Intonation

The scientific content of your presentation should be described as simply and clearly as possible, but not too simplified. Trying to impress the audience by using complicated and highly technical language (this is especially common among representatives of the humanities) is often doomed to failure and completely unsuitable for communicating important ideas. Whatever Einstein's aspirations, you should try to express the complicated in as simple and understandable a way as possible. Therefore, in a scientific English lecture, you should try to convey only one idea per sentence and, if possible, use no more than 15–20 words per sentence. Remember, the audience cannot hear a spoken sentence a second time, just as you cannot read a misunderstood sentence repeatedly in a publication.

This aspect of having only one chance in a presentation could be considered a disadvantage compared to a publication. But there are also inherent advantages. Listening to and watching a presentation has an added value compared to reading a publication: the presenter can better extract the essence and convey the crucial message to the audience. In this way, life can be breathed into the message. However, this can only be achieved if the presenter places the emphasis and makes full use of the stylistic devices discussed above. Make extensive use of these, and do not be afraid to repeat an important brief statement occasionally so that it is better internalized. You should also make such comments as "this is the most essential part of my talk" to give clear emphasis. However, do not overdo this. If every slide has such an "essential point", then there is a danger that everything will end up being equally forgotten.

If you're presenting in a foreign language. you may need to do some work on your intonation. A good understanding of written English is no guarantee of correct intonation. When in doubt, ask a native speaker or colleague who speaks the language better than you. At a conference I attended, a French scientist kept saying "and we eat the samples." Only after a few repetitions to the amazement of the audience did it become clear: he meant *heat* and just couldn't pronounce "h". So the silicon samples were not eaten, but heated.

No one will criticize a non-native speaker criticize for the fact that his vocabulary, sentence structure and intonation can not compete with the perfection of a native speaker. The main goal is to express the content clearly. Therefore, for the benefit of non-native speakers, do not necessarily try to include overly complicated and little-used expressions in your presentation just to show what a complex foreign language you speak. Keep your messages short and easy to absorb.Express the same idea again in different words, and your audience will understand the content better.

Be aware: The international language of science is broken English. Accordingly, at a major international conference, a professor from Oxford, where standard English is spoken along with Cambridge, began his talk with the opening words, "Ladies and gentlemen, I apologize for not being able to give my talk in the international language of science – broken English."

Had he even made use of the full vocabulary of the Oxford dictionary, his lecture would have been much more difficult for non-native speakers to understand than the lectures of colleagues from other countries where English is not the native language.

To conclude this discussion on language I would like to point out a common bad habit that affects native and non-native speakers alike: the frequent repetition of meaningless phrases like "if you see what I mean…", a frequent affirmative "okay?" at the end of a sentence, any meaningless repetitions, a constant "uh". Your language should be a means of communication. not ridiculous. The more questionable your speaking style, the more the transmission of information to the audience suffers. But such bad speech habits are usually unconscious and unintentional. To remedy such weaknesses, criticism from colleagues and friends or watching your own presentation video can be helpful. can be helpful.

Keep to the Lecture Time

Although you have already gained some experience, it may happen that you run into time problems during your presentation (perhaps you have already had to start late, or you have been asked to shorten your presentation). In this situation, the biggest mistake would be to speak faster in order to present everything in less time. Then nothing would be understood at all; and you would convey less information instead of more. In this case, it is much better to continue speaking at a normal pace, but to omit some prepared material, preferably such points as you had already marked in your mind for such contingencies. You should have more than one option ready to end your talk, as discussed above. Omit supplementary material but never the conclusions. These are important for people to remember your talk later.

2.4.2 Positive Feedback to the Audience

Sure you have already experienced that you give a better presentation if you succeed in establishing a positive relationship with the audience. with the audience. Then the presentation is much more relaxed and lively. If you have a positive attitude towards the audience, it will be easier for them to develop such a positive relationship with you.

You have probably also observed in other presentations that a presentation became better after the presenter succeeded in developing a positive feedback loop. …positive feedback. However, as is a peculiarity of psychological and emotional conditions, it is difficult to define what it takes to develop such an attitude, and it is difficult to establish infallible instructions for achieving a positive atmosphere.

In the following, I would like to give you some recommendations on how to create a positive communication climate. can be established. Basically, if you follow these guidelines and your audience is not made of stone or hostile to you for other reasons, you should succeed.

Attract Attention

The three golden rules of audience contact… are:

- Look at your audience.
- Speak to your audience.
- Respond to the audience.

If you follow these rules, you can achieve the attention and interest of the audience and create a positive atmosphere. Direct eye contact is extremely important. You can also see if everything is understood or if the participants' eyes glaze over.

In this context, it is also important to stand and not sit when speaking in front of a large audience. This way you will be seen better by everyone and you will also speak better and with a fuller voice than when sitting.

But even in standing lectures eye contact is not always made. Some presenters are constantly bent over the projector, others are busy reading out what is projected on the wall. They keep their lecture facing the wall rather than the audience, showing them only their back. This is the most typical lecture error, seen in about 70% of all lectures. If you point the laser pointer to something in the projection, then immediately turn back to the audience and speak to them and not to the wall. In extreme cases, the lecture participants have seen the speaker's face for the first time in the discussion. Therefore, when giving a talk, it is recommended that you look at the screen in front of you and not on the projection. Then always turn your face to the audience.

Lack of eye contact can also be interpreted as insecurity or arrogance. And if you look at the participants, as in Fig. 2.8, you automatically speak more clearly and louder than if you just say something to the wall for yourself.

Let's say you attend a lecture where the speaker hardly ever looks into the audience. So not only will you feel like you're not being addressed at all, but you'll also have to concentrate much harder to take in what's being said. The result is often that you no longer feel obliged to give the speaker your strained attention. You'll be looking out the window, counting the people in the room, checking the clock and your email.

Whereas in a conference presentation try to address all participants, when defending a doctorate or a project, it is advisable to focus on the decision-makers. is advisable.

Watch closely how the audience, who will determine the success or failure of your project, reacts to what you present. Be eloquent enough to turn

Fig. 2.8 Convincing audience contact

frowns, misunderstandings or rejections into agreement by providing additional or better explanation. Address the decision-makers directly and try to capture and successfully respond to their behavior.

Keep Attention

There are two main reasons for attention waning.: Either the audience is already sleepy before your talk because the previous talk was so tiring, the session is too long and tiring; or the speaker puts the audience to sleep with his boring style of delivery. The most critical time is after lunch, when digestion requires the most energy, or in the morning after the lavish conference dinner that spills out until after midnight. The ability to concentrate varies with the course of the day, with the low point being in the early afternoon.

So if you're looking for an attentive audience…try to get your talk into the program for the morning or later afternoon. But if you want to avoid too many critical questions, perhaps when defending a weak dissertation, choose the time immediately after lunch. If the critics are sleepy, they will ask fewer questions. Most of the time, however, you will have little ability to influence the presentation time on the time of the presentation. What should you do if your audience's eyelids drop, they seem sleepy, and they start yawning? Then wake them up! Open the windows so that oxygen-rich air comes into the room again, tell a joke or a funny story, mention interesting side issues, bring a personal touch into the lecture. And after the people are awake again, speak

Fig. 2.9 Tired lecture participant © hoozone/Getty Images/iStock

a little more emphatically and in such a way that the attention does not drift away again.

If you are in the unfortunate situation of meeting bored and sleepy participants after a tiring previous talk (Fig. 2.9), make it clear that your talk has much more life and is much more interesting. Choose an introduction that is focused on the audiencethat will generate interest and possibly hilarity. Emphasize expressively, speak in such a way that everyone can understand everything, enjoy listening to you and awaken from lethargy.

If you notice during your presentation that some participants look uncomprehending, then react immediately: Repeat what may have been a too-quick explanation in slightly different, more detailed, slower, and louder words. Break up statements that are too complex or too difficult to understand into smaller, easier to understand blocks.

In lectures and talks, a certain interaction with the audience can be very effective in loosening up the audience, entertaining them and stimulating their interest. Sect. 2.4.2 focuses on these aspects for longer presentations.

2.4.3 Keep the Attention of the Listeners

Dialogue with the Audience

For a seminar or presentations that are not limited in time, the dialogue form is with the audience is useful. You can already invite them to interrupt the presentation with questions in your introduction, so that they do not have to be held back until the end and thus the advancing understanding is impeded by intermediate steps that are not understood. Timely questions can lead to the opening of direct interaction between presenter and participants. If participants have the opportunity to contribute directly to the discussion, they will feel more involved and motivated and follow you more attentively than if they see themselves only as distant consumers of information. Then the participants feel that they are not just being presented with something, but are directly involved in the development of ideas, which arouses activity.

Any time pressure – either appointments, expiring room bookings or other – restricts the freedom of informal discussions. However, one should also take into account that long discussion events do not get out of hand into the immeasurable, and agree on a certain time frame beforehand. The most tireless participants can always continue the discussion afterwards at the coffee machine or in the next bar.

Interspersing a few anecdotes, jokes and humorous images in otherwise very serious material also helps to keep the audience interested and attentive. It is also always well received when personal aspects are included in a lecture. This creates interest and the audience can relate to a lot of things better. For example, tell about the story of an accidental major discovery, how your research came about, arguments between scientific competitors, controversies in interpreting the results, the involvement of famous intellectual figures, the struggle to get the seminal article on the work presented accepted in a journal. An additional personal relationship to the scientific subject that everyone can relate to always adds interest.

I would like to tell you a story about this: A few years ago I was invited to give a talk at the University of Missouri on a newly developed process in semi-conductor technology. When I had talked to a number of professors before the talk, I saw that only a minority of the later colloquium participants had any relation to this topic. In order not to give a special lecture to non-specialists, I decided at the last moment to give another lecture under the same heading, namely, on the patent and licensing battle between two large companies, which I decided on the basis of documentation of my publications, as a result of which nine of the original ten patent claims were overruled by the court. I used the originally prepared scientific material as illustration and evidence for what was said. In this way, the participants themselves absorbed scientific information outside their immediate scope of work with interest because the related litigation was so gripping. But this approach is not to be recommended as a general line. If I had made myself aware of the research spectrum of the institute beforehand, the lecture would have been prepared in a more interest-specific way.

Not everyone cuts a good figure when telling jokes and stories. Therefore, don't try to frantically insert such interludes into your presentation. A badly told or misunderstood joke is worse than none at all. As you become more experienced in giving presentations, you will develop the ease and sense of including appropriate humorous aspects. of humor. At first, you may have to make a note of where to place which joke. But later it will come naturally.

If you are forced to speak on a rather boring topic, say the ten-year history of the Institute at an anniversary, then including interesting peripheral information can make the talk a success. I experienced this once at such an MPI festive colloquium. The director gave a very impressive presentation on the development of publication activities, citation rates and international composition of the staff. In this way, an experienced lecturer can present even a seemingly unappealing topic in an entertaining way.

2.4.4   Giving Lectures

Finally, I'd like to make a comment about giving lectures… about giving a lecture. In lectures, it is essential that students understand and absorb what you are saying.

It is advisable to check towards the end of a sub-topic by means of a few test questions whether the desired understanding has been achieved. If you do not choose this method, summarise the essentials again clearly and in an understandable form.

It is also useful to discuss what has just been taught using example cases. Before moving on to the next topic, students should be asked if they have any questions about what they have just been told, in order to clarify any misunderstandings. to clarify any misunderstandings. In this way it can be achieved that the students leave the lecture with a certain basic understanding and are not only given the topics that they have to read up on in textbooks to understand, as unfortunately happens again and again in too many lectures.

2.4.5   Body Language

The Importance of Body Language

In the previous sections, we discussed two of the three channels of communication: visual communication through the information delivered via your

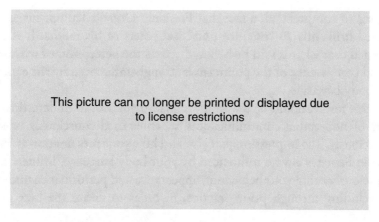

Fig. 2.10 Debate between George W. Bush and Al Gore before the US presidential election © JEFF MITCHELL/REUTERS/picture alliance

slides, including the other visual aids, and verbal communication when speaking to the audience.

In addition, there is a third essential communication channel, namely nonverbal communication, commonly called body language. commonly referred to as body language. The Chambers dictionary defines body language as "communicating information through conscious or unconscious gestures, posture, facial expressions, etc."

In this section, I will discuss the aspects of body language that significantly affect the impression created by the speaker.

Most lecturers, especially academics, will rightly say that the content of the lecture is much more important than the psychological effect of body language. But there is some surprising analysis on this, which says that this is not always the case. After the debate between Al Gore and George W. Bush before the American presidential election in 2000, psychologists asked the audience what most influenced their impression (Fig. 2.10).

The remarkable result of this survey was:

- 55% through body language,
- 38% by voice,
- and only 7% by the content.

These percentages certainly don't apply to scientific presentations.

There is an essential difference between political speeches and scientific lecturesUnlike scientific lectures, where everything must be proven, political speeches generally contain many unproven statements that are difficult for the addressee to comprehend, a fact that President Donald Trump, for example, exploited brilliantly to convince uncritical voters of his merits. It is simply demanded that what is said be believed. Here, the perception of trustworthiness and competence of the politician is strongly influenced by the expression of the body language.

But this result should be a warning signal to you not to underestimate the aspects of non-verbal communication, whether used consciously or shown unconsciously. The impression you give and how your presentation is received by the audience is always influenced by your body language. influenced. Try to consciously bring your behaviour, appearance and performance under control, including through prior practice, in order to create the best possible impression.

The most common mistakes in the conscious or unconscious use of body language are discussed below.

Typical Errors

You can easily avoid a number of the typical mistakes if you have become aware of them:

- speak to the wall (error in two-thirds of all talks, Fig. 2.11),
- obstruct the view with your own body or
- interfere with the projection by your own shadow,
- be inappropriately dressed be.

More difficult to get a handle on is a set of involuntary bad habits:

- inappropriate facial expression,
- nervous gestures or movements.

Let's deal with these problems in detail.

Speak to the Wall

At conferences, the majority of presentations are given to the wall rather than to the audience. The speaker is so engrossed in the information projected on the wall that he speaks exclusively to the wall. It is impolite to show only the

Fig. 2.11 Speaking to the wall and block the view

back to the audience, which can be perceived as showing the cold shoulder. The speaker is harder to understand unless he is using a microphone.

But the microphone is only a partial compensation. One should not underestimate the importance of reading lips to assist in capturing the spoken word. Even people who hear well have a better perception of what is being said when the speaker's face is seen. Therefore, look at the audience as much as possible and read the text better from the computer screen or projector in front of you, because you will inevitably show your face to the audience. If you have to turn to face the wall to point, still speak at least half to the audience, and only briefly, before turning fully back to the audience.

Disturb the View

Also avoid blocking the view, either by shadowing the projectionwhen standing between the projector and the screen (this happens when the projection is too low), or by blocking the view for some of the participants. for some of the participants. These are typical mistakes made by inexperienced presenters who are then asked to change position. As soon as you feel blinded by the projector, you are standing in the wrong place.

However, you should not stand too far away from the projection with your computer, otherwise the audience will have to constantly change their line of sight between the images and the speaker.

Clothes

How should you be dressed for your presentation? Although scientists give less importance to dress issues, you should try not to be dressed too casually or too formally. Jeans and a T-shirt are fine for the lab. However, they are inappropriate in a plenary lecture at a major international conference. Here, a jacket or skirt and blouse are more appropriate. However, if you were to appear like this at a talk at a company in Silicon Valley, you would look like someone from Mars. If you are unsure of what environment to expect, choose the middle ground as a compromise, neat pants and a serious shirt or sweater. Although I don't want to sound like a nagging mother, don't forget to comb your hair before you go on stage, in case there's something you need to do.

It's easy to trade in a tie for one that fits more. But it is much more difficult to get rid of some nervous habits. For example, stuttering… is not easily overcome and can only be controlled by prolonged training. But a number of other negative or nervous habits can be overcome with patience and the help of friends/colleagues or self-critical viewing of one's own video. can be stopped. These habits are:

Inappropriate Facial Expression

Avoid facial expressionsthat indicates arrogance or insecurity. But better than focusing on avoidance is to consciously smile warmly and authoritatively, which sends positive signals to the audience.which sends positive signals to the audience. When you smile, you also feel more in tune with yourself and the world.

Nervous Movements of Hands and Eyes

Thereby you will also make the lecture participants nervous. If you manage to stay calm, you will also feel calmer and less nervous. Don't constantly scan the rows of audience members with an uneasy gaze. Keep your talk calm essentially in one particular direction, as if you were speaking only to select groups of the audience.

Nervous Pacing Back and Forth

At you should move slowly and in a relaxed manner. and not constantly remain in a basic position like the guard at Buckingham Palace. But rapid and nervous running about and jumping about should be avoided at all costs. Take care where you stand, that you do not trip over power cables and steps or suddenly step out of the spotlight into the dark.

If you are unsure how you should stand, there is a simple tip for the beginning of the lecture: stand inconspicuously on your toes and drop down. This will not be noticed, but at least you will be in the appropriate position.

Gesticulating Too Much

In the use of gestures there are national peculiarities. While Asians and especially Japanese hardly use their hands to underline what is being said, this can be experienced to excess in many an Italian lecture. In northern Europe, the habits are somewhat more moderate.

However, avoid the schoolmasterly raised index finger or arms folded in front of the chest. The latter gives the impression of insecurity. If you want the audience to focus on the content of the presentation and not on the speaker's curious appearance, do not make your gestures the main source of entertainment.

Casual Appearance

Having your hand in your pocket can make you feel relaxed. In the UK, no one will mind. In continental Europe and the US, however, you will be criticized for it. Especially if you have a dissertation or project proposal to defend, show your respect for the audience and senior colleagues even by refraining from such habits. Appropriate behavior is appreciated.

Sniffing, Head Scratching, Nail Biting, Nose Picking, Etc.

All these unpleasant behaviors will give a bad impression of you. To make sure you don't unknowingly have any annoying idiosyncrasies… ask your friends.

2.4.6   Being Relaxed

Stage fright is part of a public performance, as most actors report. A slightly higher adrenaline level makes you appear more quick-witted and decisive. But too much nervousness… tends to get in the way. The secret to acting right in a speech or discussion also depends on the right level of adrenaline. As Vincent Di Salvo, a professor at the University of Nebraska, said, "Your goal should not be to get rid of the butterflies in your stomach, but to convince them to fly into formation."

Excessive nervousness is often a problem for newcomers to presenting. If you approach your presentation in a relatively relaxed manner, you will not only give a more confident impression, but also be calmer and think more clearly during your presentation (Fig. 2.12). Nervousness leads to rapid speech, less emphatic expression and forgetting important parts of the presentation.

Fig. 2.12   Convincing appearance of the speaker

In the previous discussion of body language (Sect. 2.4.4), I implicitly pointed out that there is a direct link between feeling nervous and acting nervous, with these two sides mutually reinforcing each other.

But this kind of feedback also applies to feeling relaxed and appearing relaxed.. With a little practice you can learn to use this positive feedback to your advantage. If you manage to control your behavior (facial expression, movements, relaxed posture, etc.) so that you appear relaxed, then it will help you to feel relaxed.) so that you appear relaxed, this will help you to feel relaxed as well.

Good preparation of the presentation helps a lot to prevent nervousness from arising in the first place. However, if you prepare the speech only at the last minute until well into the night, then this overtiredness will have a negative effect and will be visible to everyone. Think of the well-rested politicians at their appearances who also want to be convincing by appearing relaxed. Sufficient sleep is a prerequisite for a relaxed and spirited appearance at a lecture or discussion.

Breathing techniques can help you achieve relaxation and break nervous habits. Not only yoga teachers, but also athletes and singers benefit from the use of appropriate breathing techniques. Even medical professionals have recognized that proper breathing supports well-being. Breathing techniques have played a role in spiritual, religious and other forms of meditation and various religions for centuries, e.g. yoga and transcendental meditation. All these techniques produce a relaxing (and perhaps even transcending) effect through regular, controlled and deep breathing.

By using some simple breathing techniques, you will learn to relax better and deeper, especially in such stressful situations as giving a lecture, overcome anxiety, sleep better, get off tranquilizers, and overall gain vitality.

What is proper breathing? Correct breathing is natural breathing such as babies. It is essentially abdominal breathing. Such breathing is relatively slow, especially when combined with mid-lung and upper-lung breathing, and has a certain calming effect as it exerts an effect on the solar plexus of the nervous system in the abdomen. It also has a stimulating effect on the internal organs and provides good circulation. Through the use of abdominal breathing your voice becomes fuller and sounds deeper and more pleasant. In contrast, when you are nervous, you only breathe with the upper part of your lungs, which automatically makes your voice sound higher and thinner. Have you ever noticed that nervous people tend to speak in a high and thin falsetto voice?

If you practice this breathing technique regularly, you will speak with a fuller voice on other occasions. In tense situations, practicing this technique repeatedly will help you become relaxed and focused. These techniques also support a relaxed and healthy state of mind and body. In addition to breathing, calming and yoga techniques there is a rich selection of further literature [39, 40].

2.4.7   Pointing to the Images

Let's return to the lecture hall Let's return to the lecture hall and consider the various ways you can point to your material projected on the wall. It enhances the attention of the audience if you point to the images you have just discussed or the text you are discussing.

There are two methods to point to the projected images: Laser pointer and pointing stick.

Many presenters use a longer stick to point directly at the screen. This is a natural way to do it, and it also adds some movement to the presentation. Some colleagues feel that holding a pointing stick gives the impression of being in control of the presentation and the audience at the same time. However, when using a pointer, one should always be aware of where one stands. A right-handed person should point from the right side of the projection (Fig. 2.13). If you are standing to the left of it, you had better use your left hand. This is because if you open when pointing, it is psychologically better than closing. After each pointing, you should turn your face back to the audience.

If there is nothing to show, it is better to look at the computer rather than at the projection when giving a talk, so that you are facing the audience

Fig. 2.13   Left wrong and **right** right pointing with the stick

directly. If you are standing away from the computer and saying something about the content of the projected image or text, you should be familiar enough with the content that you do not have to keep looking at the projection to read it. If the pointing position is quite far away from the computer and you don't want to be constantly walking back and forth, you can also point to what is being discussed from your seat with the shadow of the pointer. With this technique, however, you need a strong and steady arm, as you should not show a blurred shadow and should not let the stick rest on the wall.

Alternatively, a laser pointer can be used. With it you can well point to the details and do not have to move away from your position. But if you are very nervous, it is difficult to keep the beam steady on the desired spot. In this case, a circular movement around the thing to be emphasized is better than restless fixation on a point. But there should be no constant circling about. It also happened that a speaker forgot to turn off the laser pointer again and dazzled the audience while gesticulating.

2.4.8   Time Discipline

Especially at conferences and other events with a fixed schedule, it is important that the presentations are given within the time designated in the program. If your presentation is well planned and prepared, it can easily be given within the allotted time. If some presenters exceed the time frame, there will be no discussion time. Possibly, the subsequent presentations will start late and the program will be disrupted. Therefore, colleagues who choose between parallel sessions do not reach the desired talks at the right time because they start late, which in turn causes other talks to be missed, breaks have to be shortened, or the end of the event stretches into the evening. Time discipline is especially important at conferences with several parallel sessions, so that the desired other presentations can be achieved. Exceeding the time limit you will not gain any points with the other conference topics.

Therefore, in the planning and material selection phase, you already need to pay close attention to whether you are preparing a 15- or 30-min presentation. The previous exercise will give you a feeling for the time needed. Adjust your talk to the speaking time allowed. If you are unsure during your talk how much speaking time you have left, ask the session chair. If you are behind schedule, do not rush and speak faster, but leave out parts of the speech as discussed beforehand. Always have the alternative of an abbreviated presentation ready. This means leaving out less significant results or discussions. It is recommended to plan the presentation in such a way that everything essential is presented in the first approx. 75% of the planned presentation time, followed by additional explanations. Even after about 90% of the presentation time, the option of omitting the remaining prepared 10% should be planned for. But the time for pronounced presentation of the conclusions, the main message of your presentation, which you want the participants to remember, must always remain.

Don't use up the discussion time with the presentation. The discussion can also provide you with important new ideas, because the participants will constructively deal with your new results.

2.4.9 Evaluation

At some conferences, participants are asked to fill in a checklist for the presentations. You can also make such a list yourself and ask some colleagues for suggestions for improvement (Table 2.1). If your presentation was video recorded, you can evaluate the lists and practice some self-criticism.

2.5 Surviving the Discussion

The overall impression of your presentation and the image the audience forms of you can be greatly influenced by the discussion. You can have prepared your presentation so well and delivered it competently and still ruin everything in the subsequent discussion if you do not react well and convey an incompetent or poorly informed message. if you do not react well and give an incompetent or ill-informed impression. Especially in a defense, the discussion decides the grade.

2.5.1 Answering Questions

Avoid Points of Attack

Remove anything in your presentation that is dubious or could be a point of attack. points of attack. Focus on the informative and positive and avoid the negative.

Table 2.1 Checklist for the evaluation of a presentation

(bad)	5	4	3	2	1	(perfect)
Content	£	£	£	£	£	
Structure/Organization						
Comprehensibility of content	£	£	£	£	£	
Fluid flow	£	£	£	£	£	
Information content	£	£	£	£	£	
Message	£	£	£	£	£	
(bad)	5	4	3	2	1	(perfect)
Slides	£	£	£	£	£	
Readability						
Information content	£	£	£	£	£	
Font size	£	£	£	£	£	
Color insert	£	£	£	£	£	
Animations	£	£	£	£	£	
Delivery of the speech	£	£	£	£	£	
Overall impression						
Speed	£	£	£	£	£	
Comprehensibility	£	£	£	£	£	
Intonation	£	£	£	£	£	
Emphasis	£	£	£	£	£	
Body language	£	£	£	£	£	
English quality	£	£	£	£	£	
Other comments	£	£	£	£	£	

Be Prepared

If you think about it carefully, you will see that there are a number of things you can do to make a good impression in the discussion. First of all, of course, you need to be very familiar with the material to be presented and the state of knowledge in your field. Double check that you really have fully understood all the logical steps of your explanations and any derivations. Move only on safe ground, i.e., remove anything doubtful from your presentation and do not try to explain something that you have not fully understood. Are there some weak points that provide points of attack? Then consider whether you really want to get involved in discussing such things. If so, set up a possible line of defense beforehand. …to defend yourself.

Think about possible questions and consider the answers to them. Take advantage of the opportunity to give a test presentation at your group's seminar. Ask your colleagues to ask as many questions as possible. The same questions may come from other members of the audience during the discussion.

An efficient way to get expected questions is to provoke them. How can you make this happen? If you explain something, but deliberately do not draw the conclusion expected by any attentive listener, then you can be sure to be asked about it. You can even have additional slides prepared to answer such expected questions. After all, you can say in the expected question that this picture was omitted from the lecture due to time constraints. By doing so, you can make a good impression. However, it is not advisable to make arrangements with friends and acquaintances about questions to be asked. about the questions to be asked. In this way, at least some questions could certainly be answered, but if this becomes known, the impression will be much worse than if the question is not answered comprehensively.

Respond to Questions

If a question has not been asked with the microphone, it is advisable to repeat it again aloud so that all participants in the presentation know what is being answered. Even in a smaller room, a question may not be understood from the front row at the back. In larger lecture rooms or if the questioner's language is difficult to understand, the repetition should definitely happen. By repeating the question, you gain a little more time to think about the answer. If a question is unclear or phrased in a rather negative and aggressive way, you can modify the repetition to make it more positive and answer it in a correspondingly positive way.

Asking a counter question is an effective way to respond to an aggressive question. This forces the questioner to answer first. And usually the point of the question is already broken by then. It also gives you more time to think about a suitable answer (Fig. 2.14).

Most questions are usually simply for clarification or additional information about the work presented and are easy to answer if you are well prepared and not too nervous. But some questions, especially those that go well beyond the field you are familiar with, are not so easy to answer. If you are lucky, you happened to read something about it once. If you can't give a straight answer, it's still better to give a few explanations than to stand there speechless. than to be left speechless. Think of politician interviews where the reporter asks a question and the interviewee makes a comment on a completely different topic. This is not a good example by any means, but it is a possible way out to get out of a jam in an exceptional case. You can then only hope that the questioner will not continue to insist on the critical subject matter. If you have done your work in cooperation with other colleagues and a question comes up about the technique with which you are not at all familiar, you can also pass the question on to your present co-author without losing face.

Fig. 2.14 Defending against aggressive questions (Courtesy of Claus Grupen)

If you really can't say anything, it's often best to just admit it with words like, "This is an interesting consideration. I will think about it." Or, "This is outside my expertise, but perhaps you can comment on it."

However, if everyone expects you to be able to answer a certain question asked, but you have a "blackout" or have never thought about that obvious question and don't want to show it, what should you do in such a situation? For example, you might look at your watch and say, "Answering your complex question will require more time than the remaining three minutes of discussion time will allow. Can we talk about it during the break?"

Then this colleague will be the only one who learns that you know nothing at all about it. And if you're lucky, you'll even be able to avoid meeting this colleague again... Stop.

Questions that are obvious nonsense, incomprehensible or incomprehensibly presented are also incriminating. Please do not try to embarrass the questioner for the absurdity of the question or his poor English. Tell something that you consider obvious. Such a questioner is often already satisfied with that.

Always try to develop a positive attitude towards questioners. Don't react arrogantly if a question comes at the beginning of the presentation and you know that the questioner has spent it in a deep sleep or is extremely obtuse. Rather than condescending with comments such as "as I have already explained in detail, but I will repeat it for you …", say "this is a very interesting question" and repeat the explanation in different words.

Also be aware that not all questions are asked out of pure interest. Sometimes the questioner just wants to put himself in the spotlight and show that he also knows a lot in this field, or to draw attention to his presence in front of a larger audience. In this case, the answer is of secondary importance and should only sound reasonable, confident and competent.

Fortunately, most of the questions are reasonable and rather harmless. If you feel well at home in your field, answering questions should not be a serious problem for you.

Interrupting Questions

It doesn't throw you off if a short question comes in a lecture that only needs a short answer. After the short intermediate explanation, you can continue. But don't let your lecture be ruined by long questions that require long answers. You will then lack this time to present the material prepared exactly for the lecture time. In such a case, it is better to say, "I would like to answer this question in detail at the end of the lecture," and continue.

2.5.2 The Art of Asking Questions

Most often, questions are asked out of an interest in discussing the results presented in more detail. If such a question occurs to you while you are attending a talk as an audience member, don't be too shy to ask your question. your question. The answer could be useful for your own work and help move the discussion and interpretation forward. Bad questions are quickly forgotten. The only thing that will remain is that you also participated in the discussion. You know the saying: "Any kind of publicity is good publicity".

Some scientists, usually one or two at each conference session, place a decidedly high value on their publicity. They regard the discussion time as an opportunity to present themselves, even if they have not been granted a lecture. In the 1970s, at an evening meeting of the German Physical Society, there was an amusing evening talk by Jose Liebertz (University of Bonn) on "Scientific Dialogology.". In it he classified the suspect methods of asking questions. For those of you who, by hook or by crook, always want to ask a question, the cynical guide is added to the following list.

Examining Question

If a participant is convinced that he knows much more about the topic than the presenter, the examining method is used. The aim is to raise one's own level by lowering the level of one's colleague.

Skeptic Question

This technique is often used by older and more distinguished colleagues when younger colleagues are not allowed to succeed. The goal is usually to cast doubt on the results or conclusions presented. This method only works from the top down. If an experienced scientist declares not to be able to comprehend what the younger colleague presents, it means it was nonsense. Conversely, if a young scientist remarks that he can't follow what the high-profile colleague is saying, then that just shows how limited his scientific understanding is.

Deliberate Misunderstanding

This method does not necessarily give the most attentive impression of the questioner, but in order to appear at all, it is occasionally used. Such situations include: If a presenter said that results were obtained only at normal pressure, the question might be, "If I understood you correctly, you obtain your results only at positive pressure. But our measurements show such effects only at normal pressure. How do you explain the contradiction?" The speaker will of course agree with you, but you have set the scene.

Method of Modified Boundary Conditions

This highly effective method you can use even if you have not understood anything at all. Only a few details need to have been picked up. For example, if the experiments presented were performed at 500 °C, you can simply ask: "And what do you expect at 600 °C?"

Method of Autapotheosis or Self-Congratulation

Here the questioner tries to make clear that he is in a higher league than the speaker. Introductions to this can be: the reference to the questioner's highly cited article in the journal *Nature,* his recent conversations with close friends of Nobel laureates, or the reference to his highly regarded plenary lecture at a much larger international conference.

Method of Deviation

Here the questioner leads the discussion to a fringe area of the lecture in which he is a much more proven expert, hooks on to an aside to show that he knows much more here.

Prepared Questions

Few People come to a conference with a firm desire to ask a question, but are not sure if they will actually think of something during the talk. Based on the abstract, a question is then thought of and written down. At a conference where the president of the American Physical Society was giving a talk, one attendee stood up and said, "I have a question." He then began to read out a four-page manuscript, concluding by remarking, "Explain to me why this excellent article was not accepted in the journal *Physical Review Letters!*"

Stupid Question

At You should be alarmed when someone starts by saying, "I have a stupid question…". This type of question is usually asked by colleagues who are not at home in the field in question, but who have a broad overview overall and think deeply. Perhaps this colleague has just pinpointed the weak point in your argument. Such questions are usually anything but stupid. If the speaker can't answer them, he is the stupid one.

2.6   Poster Presentations

2.6.1   Classification as a Poster Lecture

Often a scientist is disappointed if he or she is only granted a poster and not an oral presentation because of the excellent abstract submitted. Others are relieved when they are not subjected to the stress of an oral presentation. Posters are often perceived as inferior conference contributions.

But a poster contribution also has something positive in comparison to an oral presentation: You have more time to discuss your results with really interested colleagues before the poster than in the short discussion time after a presentation. You can thus get new ideas for your own work and perhaps also develop new collaborations.

Many conferences get so many good abstracts offered that the conference time simply does not allow the oral presentation of all the good contributions. Therefore, the classification as a poster is not a judgement of quality, mostly contributions on more specific topics are affected.

2.6.2   Visual Considerations for Poster Design

Try Make your poster stand out among a few hundred other competing posters. This is less a matter of content and more a matter of visual design.

The first and often only part read by colleagues strolling through the rows of posters is the title. But even this will not stand out if it is difficult to read, printed too small, too long or too technical. That's why it's important to choose a title that is concise, as short as possible, and in large print. A question as a title will also create interest. Don't forget to include your name and address underneath. The abstract, which is read next with an appealing title, should be short and meaningful. The body of the poster that follows should contain as little text as possible and be well structured with subheadings. As a guide, have no more than one A4 page of text (font size 14 point) in total. The text should outline the experiment, include only the main aspects and a summary of the work and the conclusions, but not a detailed discussion. Minimize the experiment or theory description. The results should also be reduced to the most important aspects. If your results lead to further conclusions, other colleagues working in the field will be interested in them. Choose at least font size 14 point, so that your poster text is still readable from a distance of 2 m in case of a large crowd.

Focus especially on the illustrations. In them, your results will become clear. The ratio between text and illustrations should be balanced. The illustrations must be as self-explanatory as possible and not too complicated to understand. The size of the illustrations must be appropriate, preferably DIN A4 format.

As with a publication, a poster should include the main citations and an acknowledgement at the end.

The most commonly used poster format is DIN A0 (84 cm × 119 cm). If possible, the entire poster should be printed in this format. If such a large printer is not available at your institute, going to the copy shop is worth the better appearance of the poster. This poster style is better than stapling 16 A4 sheets.

At many conferences there is a prize for the best poster. The prize may be a bottle of champagne or US$100–400. Try to get the prize for the best poster by presenting a poster with excellent content and exceptionally good visual appearance (Fig. 2.15).

2.7   Some Tips for Chairing Meetings

It …an honor to preside over a conference session… to be allowed to chair a conference session. This means that you have gained a good reputation as an internationally recognized specialist in the relevant field. At the same time, it means that you are considered to have the overview of the field, which is important for the selection of the invited keynote lectures and the appropriate representation of the field in the other lectures, to have good international connections and to be diplomatic enough to lead the discussion. In order not to disappoint these expectations, you should take into account the following rules:

2.7.1   Tasks of the Discussion Leader

Preparation

Concentrate When selecting the invited keynote lectures, do not concentrate on your friends and good acquaintances alone. I once had the embarrassing experience that the session chair had exclusively given keynote speeches to his research partners, who, however, were not highly competent, which then led

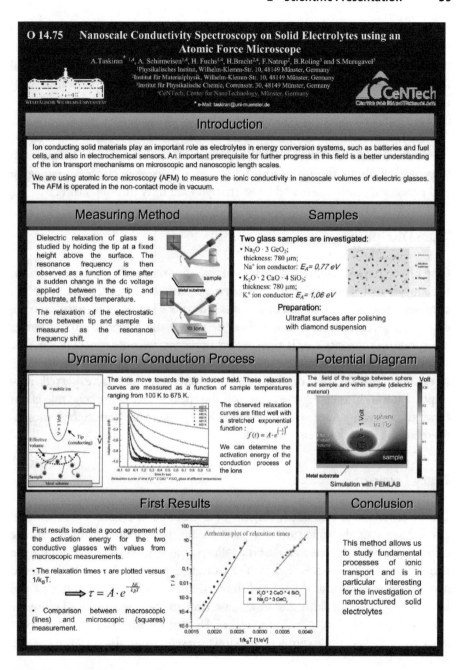

Fig. 2.15 Example of a good poster

in the discussion to the fact that he had to answer a large part of the questions put to them himself.

Before the session starts, familiarize yourself again with the schedule and the possible different length of the presentations. Nothing is more embarrassing than asking a keynote speaker who has 30 min to finish after 15 min because a shorter contribution was expected by mistake.

Can you pronounce all the speakers' names pronounce them correctly? A few introductory remarks about the content of the session and before each presentation about the upcoming topic and the speaker create an engaging and expectant atmosphere and show that you are fully in the picture.

Time Discipline

One of your most essential tasks is to make sure that each presenter starts at the announced time and does not overrun. This is especially important in conferences with several parallel sessions. For this purpose, it is useful to use a clock that gives a signal 2 min before the end of the presentation and when the speaking time expires. If the speaker overruns, you must decide whether to let them continue speaking into the discussion time, or ask them to come to the end. This decision depends on how many questions are expected, how important and interesting the topic and the speaker are. If the schedule is out of whack, you may need to ask the following speakers to shorten their presentations a bit to get back on schedule.

Fancy Lectures

If a presentation is cancelled, this is not a problem at a conference without parallel sessions. In that case, the next lecture is simply continued. In the other case it has to be ensured that the following lectures start at the announced time. This can be ensured before the cancelled presentation (which should actually become clear at the start of the session if a speaker does not show up) by granting more speaking and discussion time to the presentations before the cancelled presentation, e.g. 20 instead of 15 min. In addition, one can come back to cancelled discussions of previous presentations. In extreme cases, a short break should be taken.

Discussion Leader

The second main responsibility is to lead the discussion. After the call for questions has been made, the discussion leader should allow the questioners to speak in the order in which they reported. He/she should allow each question to be answered to the satisfaction of the questioner, audience, and speaker. If the situation arises that no question is asked, then the moderator should break the ice with a first question, even though this may remain the only question, but this will at least avoid the embarrassing situation of no question at all. It is definitely advisable as a discussion leader to think about one or two possible questions during the presentation. Conversely, if many more questioners come forward than the discussion time allows, then priority should be given to calling on those colleagues who had not previously asked any questions. If the end of the discussion time is then approaching without all questions having been presented, you as the discussion leader can suggest that the discussion be continued during the next coffee break.

2.7.2 Seminars and Internal Meetings

In internal meetings the organizer and the discussion leader are often the same. When planning, it should be taken into account that there are no overlaps with other events and that essential colleagues can participate. Preparation also includes booking a room, drawing up a plan for the content of the meeting, giving advance notice of the agenda to all colleagues involved, and possibly ordering coffee. When the meeting takes place, the atmosphere is usually quite relaxed. There is less time pressure here, but it is advisable to agree on a time frame at the start so that the event does not get out of hand. This includes agreeing on the length of each contribution. As a general time frame for internal meetings, aim for 90 min, the typical academic time unit.

The discussion leader must have the courage to break off long presentations that lack content and to continue discussions that do not concern all participants in a smaller circle. If a meeting goes on too long, it can also be officially declared over, and those colleagues who are keen to continue the discussion can do so in the nearest restaurant.

2.8   What to Do and What Not to Do

DO	Adjust to your audience.
DON'T	Don't give a specialized talk to a general audience.
DO	Select the appropriate amount of material for the allowed presentation time.
DON'T	Do not try to show all the details.
DO	Plan 1–2 min per slide.
DON'T	Do not include text or illustrations that are not discussed.
DO	Structure the presentation logically (like a publication).
DON'T	Don't jump back and forth from topic to topic.
DO	Have only one main topic per slide.
DON'T	Don't overload the slides with information.
DO	Have only keyword-like text in the slides as a lecture framework.
DON'T	Avoid reproducing complete sentences or long sections of text.
DO	Make the slides attractive by using color.
DON'T	Don't have too colorful or only black and white slides.
DO	Acknowledge those who have supported your work.
DON'T	Refrain from any plagiarism.
DO	Draw clear conclusions, and give a take-home message.
DON'T	Don't end the lecture with wild speculation.
DO	Practice the presentation with colleagues, friends or alone.
DON'T	Avoid preparing the talk until the last night before giving it.
DO	Take criticism from colleagues seriously.
DON'T	Don't keep repeating the same mistakes.
DO	Speak freely, naturally and in a friendly manner.
DON'T	Never read off the lecture.
DO	Speak clearly and check for correct English pronunciation.
DON'T	Do not recite too quickly, too quietly, without intonation, or with incorrect grammar and pronunciation.
DO	Appear appropriately dressed and behave appropriately for the situation.
DON'T	Do not block the view with your own body or shadow.
DO	Feel confident and relaxed and perform accordingly.
DON'T	Refrain from arrogant or nervous behavior.
DO	Give the presentation in the allotted time.
DON'T	Don't ignore the meeting leader's request to come to the end.
DO	Prepare for the discussion beforehand.
DON'T	Don't answer a question before the entire audience understands it.

I hope that the advice given in this chapter will be useful to you and help you to give successful presentations.

3

Publishing Scientific Articles

In this chapter we will look at how to plan, write and submit a scientific article. If you are inexperienced in this, useful recommendations will be given here. And for those who already have experience of publication, this summarized presentation of right and wrong procedures will be useful to improve writing skills.

Although I want to focus here more on writing manuscripts for journal articles rather than books, you will see that once you have acquired the skills to write good articles, you have already gone a significant distance toward tackling good teaching material and even a book manuscript.

3.1 Planning and Preparation

Mistakes in preparing represent preparing to make a mistake (Mike Murdock).

3.1.1 Before You Start

Before you start planning a publication, ask yourself the following critical questions:

Do I have enough substantial results? Check the value of your results thoroughly and critically to see if they are really worth publishing. The questions are: Are the results reproducible? reproducible? Do they lead to significant findings? Are these results interesting for other scientists? If you can answer "yes" to all of these questions, then publication is probably justified. However,

© The Author(s), under exclusive license to Springer-Verlag GmbH, DE, part of Springer Nature 2023
C. Ascheron, *Scientific publishing and presentation*,
https://doi.org/10.1007/978-3-662-66404-9_3

if your results contain little that is new or require further investigation before they can be interpreted, then it is better to wait until you have more substantial results. If you are unsure about this decision, then more experienced colleagues or your boss can certainly help you. Also, discuss your work with these colleagues in light of what other researchers have published in the field. This discussion will help you to see whether your work actually fills a gap, helps to clarify open questions or even opens up a completely new research direction.

What do I want to report? A publication is usually narrowly thematic, contains one or a few significant findings, and gives the reader a message. You should select for the content of your publication only those results that are needed for the message. Overloading the article with other, less relevant results tends to lead to a scientific dilution effect and should be avoided. Your article should focus on the topic as much as possible. Then the reader will also appreciate this article and find it remarkable.

When should the results be published? Sometimes there are good reasons for delaying publication. As stated earlier, it is important to have accumulated enough substantial results and to have reproduced them. Especially if your results are also technologically relevant, then rapid publishing is not advisable. Because in this case you should first. In Europe and Japan, the patent application must precede the publication of the results, because everything published is classified as public knowledge that can no longer be patented.

Am I under competitive pressure? When different researchers or research groups are working on the same topic, competition can develop to solve certain problems first and to have the corresponding first publications on them in order to establish priorities. to establish priorities. This is because the priority is determined by the submission date of the article.

If you are in competition with other groups and want to ensure that the newly discovered effect is named after you, you should publish your main results as soon as possible to secure priority rights for yourself and your co-authors. For rapid publication, a *short note* or a *letter is* most suitable. Because such publications are subject to strict scope restrictions, they can only report the new findings without discussing them in detail. However, if you are not under competitive pressure, it may be more appropriate to present a more extensive and in-depth discussion of the results in a longer original paper. in a longer original paper. It is a shame to waste excellent results in a hastily written publication of inferior value, as they should only be published once. To write a more comprehensive publication, you will need to invest more time in gathering all the data needed to complete the picture, reviewing the literature extensively, and thinking deeply about interpretation. Ultimately, a more informed scientific paper will have a greater impact than a terse description of results.

However, it is possible to wait too long: Carbon nanotubes were discovered by Dr. Iijima from NEC Tsukuba in 1990. But he waited 2 years to publish it, which was 2 years too long. Although this article received great recognition by appearing in *Nature*, it was too late to be considered for the Nobel Prize for New Carbon Modifications. It was awarded solely for the related fullerenes.

If you are late, others will have already taken credit and there will be less interest in your work, even if it is different in some respects from the articles already published.

Sometimes, however, publishing too early is just as damaging as publishing too late. Because it can also happen that your work is way ahead of its time. Perhaps your results only become technologically relevant much later. Then it can happen that later nobody looks for much earlier publications in this field and the work is not cited at all, as it happened to me with a newly developed semiconductor technology. There is no guarantee that pioneering work that has gone unnoticed will be rediscovered later. Work in the mainstream of research is much more likely to be cited. So you just need a good nose to be in the right place at the right time with your research.

What is already known? Although you will probably have read up on what is already out there in your field before you started your own research, an even more extensive study of the literature will be necessary when it comes to interpreting, publishing and fully placing your findings in the context of what has already been published. Other publications may provide interpretive support or contradict your own view of things. However, you should at least mention both.

Which language? Your publication will have a greater impact and receive more attention if it is published in an international English-language journal. international English-language journal. If English is not your native language, it may be more costly than in your own language. But if your results are substantial, by all means put in that extra effort to make them known internationally. How painful it can be to publish in the wrong language was experienced by Eiji Osawa, Hokkaido University, experienced.

He was the first scientist to theoretically predict the existence of fullerenes. However, he had published this work only in a Japanese-language journal (Fig. 3.1) and thus did not achieve international fame. Thus, his work was also unknown to the Nobel Prize Committee when it decided on the award of the prize to the three experimenters who demonstrated fullerenes. When the decision was announced, he immediately sent his publication, which was 20 years older, to the Nobel Committee. But he was told:

Fullerenes

First paper

Eiji Osawa, Kagaku **25**, 854-863 (1970).

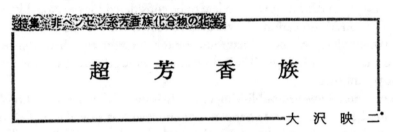

超　芳　香　族

大　沢　映　二

1. はじめに

最近数年間の非ベンゼン系芳香族化学の発展の速さと広がりの大きさにはまことに目をみはるものがある．その原動力が Hückel の分子軌道法に基づく (4n+2)π 則と，合成化学技術の近代化の二つであることについてはまず異論のないところであろう．とすると数ある目ざましい新芳香族化合物のうちでシクロプロペニウムとアンニュレンの発見は実に意義が深い．しかし逆に考えると Hückel 則の合成化学的検討に関して，最も劇的な開拓期はすでに終わりを告げ，今後は理論，合成両面にわたる精密化と誘導体化学の時代であるとみることもできるであろう．

今後の芳香族性に関する化学がさらに興味深い展開を示すであろうことは疑いないが，ここで Hückel 則を離れてまったく新しい芳香族性の出現する余地はないかどうかもう一度考えてみることにしよう．

一つの考え方として「次元の拡大」をあげることがで

• OSAWA Eiji i　北海道大学助教授(理学部)　工博

化学　第25巻　第9号

2-3 corannulene

多数のベンゼン環の縮合した型のいわゆる“縮合多環式芳香族炭化水素”は典型的な平面分子である．これらの代表的なベンゼノイド芳香族が球状分子の型をとったなら超芳香族性を示さないだろうか？　たとえばサッカーの公式ボールの表面に描かれている幾何模様を思い浮かべてみよう．それは正多面体として cube のつぎに小さな正二十面体 (eicosahedron) (12)の頂点を全部切り落として正五角形を出したもので，truncated eicosahedron とでも称されるべき美しい多面体である(13)．図ではわかりにくいところもあるので，もし手もとにサッカーボールがあれば手にとってながめていただくとはっきりするが，五角形(黒く塗ってある)の間には規則正しく六角形がうずまっている．一見これらの成分多角形はたいして曲がってもいないし，各辺はすべて同じ長さにすることができる．もしこれらの頂点を全部 sp^2 炭素

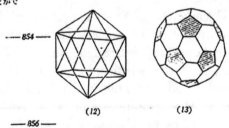

(12)　　　　(13)

— 854 —

— 856 —

Fig. 3.1 Osawa's publication on fullerenes in the Japanese journal Kagaku 25,843 (1970)

We regret that your work was not previously available to us. The Nobel Prize can only be awarded to a maximum of three people. We can't add your name as a fourth. And the Nobel Committee's decisions are never revised.

Thus, Osawa eventually missed out on the Nobel Prize because he had published in the wrong language [32].

3.1.2 Select a Magazine

Before you start writing the article, you should consider to which journal you want to submit it, because each journal has its own style, thematic orientation and author guidelines. The likelihood that your article will be accepted increases if it is written in the appropriate style from the beginning.

New or established journal? If you choose a new or smaller journal that has yet to solicit articles, the likelihood of acceptance is higher. But such journals have a lower circulation. Consequently, your article will not be read, noticed, and cited as often.

An acquaintance of the author was persuaded to publish an article in a newly founded journal. He was pleased to find that this was the most cited article in that journal, until he finally spoke to an editor of the leading physics journal *Physical Review Letters,* who told him, "If you had published your article in our journal, it would have been cited ten times more often."

Rejection Rate

If You send your article to a leading and highly cited journal, then you have to factor in that it will not be accepted. These journals are offered many more articles than they can publish. The leading physics journals, such as *Physical Review* or *European Physical Journal,* only accept about 30% of the articles submitted for publication. In *Nature* or *Science*, it is even only about 5%. And even if the article is accepted, you may be asked to make extensive corrections. But be aware: even the leading journals need to fill their pages.

Other Factors

The decision between *short note* or original work narrows down the possible range of journals, as some journals publish only *short notes* or only original work, while there are also journals that publish a mixture of both categories plus review articles. publish.

Another decision to make is whether you want to get your article published in a specialized scientific journal, such as *Progress in Statistical Hydrodynamics,* in a journal that covers the entire spectrum of a discipline, such as *Physical Review Letters* for physics, or even think about a journal that covers the entire field of science, such as *Nature* or *Science*. However, you should only consider the latter general journals if your results radiate far beyond your narrower research area and are of broad interest.

Open Access Journals
Since around 2005, the way has been opened for articles to be accessible to all interested readers in open access-journals and not only at the institutions that subscribe to the pay journals. Many scientists and research-funding institutions are very interested in making the results available to an even wider circle of interested parties in this way. Such journals exist in all fields of research.

The submission and review of articles through *peer reviewing* is carried out in exactly the same way as in conventional journals.

Initially, there were efforts to have scientific societies publish these online journals. publish these online journals. However, since the effort and costs involved in publishing open access journals are comparable to those of conventional journals, the only way forward in the long term was to leave this field to the publishers. In order to cover the production costs, it is not libraries that have to pay for this, as is the case with conventional journals, but the publishing authors or their institutions. The publication costs per article are around 1000–3000 EUR depending on the journal. For this reason, publication costs are often included in the awarding of funding for scientific projects.

However, as these funds are not available everywhere, most articles are still submitted to and published in the established other journals.

3.1.3 Citation Frequency and *Impact Factor*

An essential criterion for the selection of a journal for publishing one's own article is the citation frequency. The average citation frequency per article is called *impact factor* is called. While the *impact factor of Nature* and *Science* was 20–30 depending on the year, it is generally below 10 for the other leading journals. In order to achieve the greatest possible awareness of one's own work, the aim is therefore to publish in the journals with the highest *impact factor*.

The so-called Hirsch factor or H-index serves as a criterion for the publication success of a scientist. It is determined by plotting the citation frequency of the publications as in Fig. 3.2. The resulting weighted mean value, in which the citation frequency and the number of the paper are identical, is called the Hirsch factor.

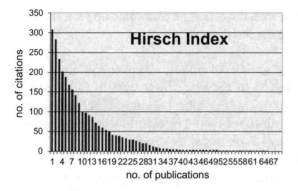

Fig. 3.2 Determination of the Hirsch factor

For the average scientist, this is 20–30, while outstanding researchers with groundbreaking work exceed 100.

In some countries and research institutions, the Hirsch factor is used as another evaluation system of publications to assess the scientific value of researchers' work and is also a criterion for awarding research grants, bonuses, professorships and advancement on the *tenure track*. In China, for example, the *impact factors of the* journals of the individual articles per researcher are added up for this purpose.

This strongly stimulates the publishing of as many articles as possible in the leading scientific journals. All scientists must follow the slogan *"publish or perish"* … which means publish or perish.

3.1.4 Who Should Be Listed as Authors?

Number of Authors

Fundamental anyone who has made a significant contribution to obtaining the results to be presented should also appear as the author of the publication in question. This concerns both the obtaining of experimental results and their interpretation. Sometimes the decision is borderline and not so easy to make. What about the person who came up with the idea for the experiment or the calculations but was ultimately not involved in it himself, or the technician without whom the experiment would not have gotten off the ground? Should the head of the research group, who organized the project funds… organized the project funds and was always helpfully available for discussion, become a co-author, even if he was not directly involved in the work? Or what should be done with the colleague who came up with the idea for the decisive solution during the discussion?

Normally, the number of authors is kept to a minimum, namely those who have been involved in the work from start to finish and who can explain everything if necessary. The contribution of other colleagues who have made any minor contributions should be mentioned at the end of the article in the *(acknowledgement) at the end of the article.*

In some institutions there are other rules. There, the boss is always co-author of all publications of his group or institute. But there are also bosses with a more scientifically correct approach who say they do not want to appear as co-authors of publications to which they have not made a substantial contribution. Such substantial contributions may be organizing the project and participating in the interpretation of the results. For some projects with many collaborators, such as large nuclear physics collaborations among numerous groups, the list of authors may include as many as 100 names. The decision as to which of these should eventually receive the Nobel Prize is then not always straightforward, as in the case of the Higgs boson, for example.

For some projects, it is a condition that the project director must appear as a co-author in order for the publication to be accepted for the project. And if you have legitimate hopes for a project extension, it is advisable to follow these rules.

Author Order

After has been decided on the number of authors of the publication, the determination of their order can become a delicate question. It is a widely accepted fair rule that the person who has the major contribution to the outcome of the publication should also be named as first author is named. This decision involves the question of how each contribution should be assessed. What should be judged as the main contribution – the original idea for the experiment, the major contribution to the conduct of the experiments, the interpretation of apparently contradictory results, the coherent presentation of results in writing the publication? Since these contributions are often spread across different colleagues, it is not uncommon for not everyone to be happy with the order of authorship that is ultimately determined. If the authors cannot agree among themselves, this decision must sometimes be left to an outsider recognized as an arbiter, e.g. the head of the group, regardless of whether he is a co-author himself.

In some institutions, the rules are quite different. There, the boss is always first authorregardless of his or her contribution. Elsewhere, colleagues occasionally make use of their disciplinary power and act as first authors or co-authors. To this end, publications are sometimes written by the more experienced co-authors. However, I also have friends who are directors of larger institutes and say:

> I rewrite all the publications of my doctoral students, but never make myself the first or second author, always the last. The main work is the production of the scientific results and not their presentation. Therefore, I leave the main credit to the one who did the main work. It doesn't matter that insiders might recognize who wrote the article.

This is a conducive behavior for young scientists because it gives them full credit for their own work.

In some institutions, where it has often proved difficult to adequately balance the contributions of individual co-authors, it has been decided to simply list the authors in alphabetical order. But this can degenerate to the detriment of the colleague whose name is at the end of the alphabet and who has done the main work. This is because if a publication has more than three authors, it is often cited with only the first author's name, and then it says et al. Thus, the key author's name may be anonymously lost and hardly associated with his or her work when the citation is read. Totally exasperated by this practice, an ambitious associate professor at the University of California LA pulled the emergency brake. She had the registrar's office delete the first letter of her name "Yard" and thereafter became the first author of her institute's publications with the new name "Ard".

However, there is hardly anything that can be done about the rules established at institutes. You just have to accept them; and to avoid problems with them, you should try to minimize the number of co-authors. If the number of co-authors is low and the contributions are clearly delimited, one can then try to arrange the order of authors according to the importance of the contributions.

Ghost Writers

It is not uncommon for a distinguished scientist to be asked by a younger colleague to act as a co-author, even if he or she has nothing whatsoever to do with the work to be presented, in order to give the publication more weight. Adding a big name is done with the intention of increasing the likelihood of the article being accepted by a leading journal. But should you go down this route, it is an unavoidable requirement that the colleague in question has at least been informed and agreed to this beforehand. Any colleague named as a co-author named as co-author must have read and accepted the paper before it is submitted. Or would you like to read in the next issue of the journal, "I object to being named as co-author of this dubious paper that I did not know beforehand." Therefore, such a practice is discouraged.

Also, occasionally certain deals are made for someone to become co-author, first author, or not author of a publication at all, thus transferring all the credit to other colleagues. The most extreme case I know of scientific recognition being ceded is that of a Russian doctoral student who was asked to hand over his finished doctorate to the son of a minister, since he was such a bright fellow that he could easily produce a second doctoral thesis quickly. He was promised that the resulting loss of time in progressing through his academic career would be more than compensated for later. The following year, he was allowed to submit a "light version" of another dissertation, and he experienced a rapid promotion to professor soon thereafter.

3.1.5   Writing Multi-Author Publications

Be aware that co-authorship does not necessarily include being involved in writing the publication. Co-authorship can also be based on providing essential ideas or participating in the experiments or calculations. But each co-author must be given the opportunity to read the article and contribute his or her suggestions and requests for improvement. All co-authors must agree to the variant of the article to be submitted.

With multiple co-authors, how should the writing be organized most effectively? Is it a good idea to have everyone writing simultaneously on pre-agreed, different parts of the article? Then, when attempts are made to put these fragments together, it can come across like the Tower of Babel: It turns out that the overall work is extremely incoherent because everyone had their own style.

The most effective way is to designate a colleague, usually the key author, to write the first draft of the publication. If the article contains two distinct parts, a theoretical and an experimental part, a theorist and an experimenter can start writing at the same time. Once this draft is available, the other co-authors should revise the publication step by step. In this way, the final result is a work that appears to be of one piece and with which all participants are ultimately satisfied.

But now I am almost reaching too far ahead. After deciding on the co-authors, the content should first be planned in detail.

3.1.6 Planning and Preparing the Content of a Publication

When confronted with the task of writing a publication some people start to sweat. They do not know what to start with, devote themselves to individual data, compare them with other results, read and quote related articles, write a few lines, write something about the experiment, ask themselves enervated where they have buried the data from the control experiments … This approach can be named as the way "let's see where it will all lead".

It is much more effective to think about the approach beforehand. It is possible that the concept and perhaps even the goal will change in the process of preparation, because new ideas come to you or new results emerge. But you should always have a clear plan.

Based on my own experience and that of many friends and colleagues, I would like to recommend the path described below. It consists of several steps that should be taken in order. First of all, it must be clear what experimental or theoretical topic you want to cover. Please note that for articles that involve only text or mathematics, which is only a minority of scientific articles, the preparation procedure begins with step V.

I. Compilation of data and results

At the beginning, all examinations to be used must have been evaluated. This is the raw data you will be working with. In this process, you may become aware of what other data is missing. Fill in these gaps. It may be necessary to go back to the lab or computer to get the missing information.

II. Selection of results to be presented

From the many results and computer printouts, you may see that this is far too much material to include in a publication. In this case, the next step must be the selection of the main results.

III. Creating the images

Many experienced scientists recommend as the next step to create the figures and tables. Especially in experimental publications, the figures often contain as much or even more information than the text. Therefore, you should not underestimate the importance of the figures and not regard them solely as decorative accessories. It is always much better to present results graphically in the form of figures than to show long columns of figures in tabular form. Always try to visualize your results as much as possible.

If you want to display the results, you have to consider which parameter correlates best with which other parameter and in which display mode the effect is most meaningful: linear, semi-logarithmic or double logarithmic. the effect is most meaningful: linear, semi-logarithmic, or double logarithmic. Or are bar charts, pie charts or three-dimensional representations most appropriate? To avoid overwhelming the viewer, do not overload the illustrations with content. In general, an illustration should only express an essential connection.

It is best to prepare the figure immediately after all the data is available in the final perfect form to be used in the article. Then this part of the preparation is already completed.

What has to be considered when creating the figures? I will explain this using the example of Fig. 3.3.

- All illustrations should be clearly structured and instructive.
- In order to avoid a confusing overloading of illustration contents it is sometimes didactically better to plan several different illustrations.
- The figure should not only contain two arrows as axes, but better a frame.
- The unit lines should be drawn in the frame on all four sides, better inside than outside.
- The frame and the lines in the illustration must have an appropriate thickness, i.e. they must not be difficult to recognise because the lines are too thin (at least 0.2 mm). (at least 0.2 mm), but also not look like a mourning frame.

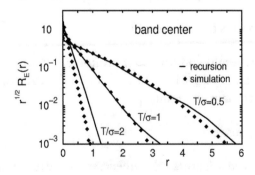

Fig. 3.3   Sample illustration for the frame around the image

- If the figure contains different curves, they should be distinguished by different line types and symbols for the data, especially if they overlap.
- The illustration must be of an appropriate size. It must not be recognizable only with a magnifying glass, like a postage stamp, but also must not appear in dimension as if it were the author's intention to scrounge for space.
- The axes must be named, either with symbols, which are to be explained in the figure text or with terms written out in full. Also, do not forget to insert numbers at the unit lines or at least add "arbitrary units" should be added. The units should be placed in round brackets, e.g. (Hz).
- In order to make text and numbers on the axes of the figure itself readable, the font size must not be less than of 9 points must not be undercut.
- Inserts are suitable for schematic illustration of experiments.are suitable for the schematic illustration of experiments.

If it is possible, it proves convenient to have a key illustration for an article. This can be a systematizing and comparative presentation of the most essential results or a derivative presentation. Many famous and groundbreaking publications have such a key figure, e.g. the Nobel Prize winning work of Klaus von Klitzing [33].

IV. Figure texts

The figure texts must be as brief and informative as possible. They should contain only one statement for understanding the content of the figure, indicate what is shown on the axes, and explain abbreviations, symbols, signs used and the lines of the various curves. Under no circumstances may they contain a discussion of the curves. This must be done in the continuous text.

V. The main message define

When you now have all the results in prepared form, you will realize what the main content of your paper is. You should now think about what the main message of your article should be. For this, it is useful to write down a first draft of the conclusions, i.e. the end of your article. Although you may want to revisit these conclusions you may modify them slightly after writing the article, it will help you to have the goal of the article clear in your mind as you write it, so that you can guide the reader specifically to the message of your article.

VI. Build the skeleton of the article

The formal structure of the publication should now be (detailed discussion in Sect. 3.3). Write a brief outline of the content and record the main issues to be discussed. Even this skeleton may change during the process of writing. However, it helps to bring logic and structure to the expositions. Additionally, write down any thoughts that arise about each part of the paper before they may slip away.

This plan should be checked with all co-authors before writing begins. This includes the selection of illustrations to be used and discussion of conclusions.

VII. Selection of literature to be cited

Perhaps you are already well informed about the state of knowledge in their field and have collected all the relevant publications of other colleagues. Then you only have to select the literature sources to be used from your rich collection. Otherwise, I advise you to do a thorough literature search at this stage, i.e. just before you start writing the paper. After all, you don't want to make an uninformed impression. Even if you have already collected a large number of relevant and citable articles, it is not a bad idea to search again in databases or the electronic versions of journals to find out what you are looking for. or electronic versions of journals to make sure that no significant articles have been overlooked. For this purpose, it is extremely helpful that the international publishers have now also retro-digitised the older issues of their journals, which were only published in paper form, and thus make them available electronically. and thus make them available electronically. You may find unexpected treasures in such a search.

What Should You Cite?
First and foremost, the foundational and perhaps already historical works that established the field should be cited. The seminal further works that ensured the progress of the field should not be forgotten. It is also useful to include review articlesthat present the entire field. This makes it easier for newcomers to get started. The next papers to discuss are those on which you are directly building. The questions left unanswered there that you will answer are to be worked out in the process. Alternatively, you may discuss conflicting papers whose contradictions you will clarify. In the discussion of your results, you will draw on related works for interpretation. Since the number of references cannot be arbitrarily high, the essential works should be cited as a matter of priority and not all possible insignificant works.

You should have read all the cited actually read all the works cited and not just from other publications. For it could be in another article the quotation discussed sense-distorting, or the name of the author, the page number or the year of publication of the article are written incorrectly. Then any connoisseur will immediately see that you never read the article and just stole the already previously incorrect citation. For this reason, many universities require that in scientific qualification papers. not only the first, but also the last page of a citation must be given.

VIII. It will be easier for you to deal with the literature sources if you have all the articles to be cited on your computer or as a printout or copy.
   IX. Writing the text

By now you have all the illustrations and literature sources to be used at your disposal. If you now arrange the illustrations in the correct order of use and place the works to be cited in between in the right places, then the core content of your publication is spread out before you. If you have now also arranged your fragments of thoughts and partial discussion drafts into the puzzle in front of you, then the picture of the entire publication will slowly rise out of the fog; and you will be surprised to find that a large part of the work is apparently already done.

The writing of the publication, which seems to most people to be tremendously difficult, will then be a surprisingly straightforward, speedy and easy process for you. Sects. 3.2 and 3.4 discuss in more detail the issues of good academic writing style, appropriate structuring of a publication, and some formal issues associated with academic publishing.

X. Revise and improve the text

After you have tried to write everything down as convincingly as possible, you may find on re-reading that you are not completely satisfied with it after all, or that a conclusive proof is still missing. This may lead to the need for rewriting or further experimentation and calculation. Or you may need to consult the scientific literature and other colleagues more extensively, which will probably result in a delay in completing the publication. However, do not let this discourage you. It is normal that when approaching a much desired goal, despite working most intensively, you often feel that the distance is not getting smaller. If you continue to work hard, you will overcome these problems as well.

It is a well-known phenomenon to encounter unexpected questions left unanswered when you try to present a problem in a coherent and convincing way, because then you are less likely to pass lightly over what seems to be known. In this way, you achieve a deeper understanding. If you are compelled to explain your findings in detail, either in an article, a paper, a lecture, or a seminar, you will quickly realize whether or not your explanations can be presented in an understandable way. Problems with this are often an indicator of incomplete knowledge of your own.

If you pursue these questions, you will often come to new insights and recognize new cross connections. Or you will come across open questions to be answered in the future. In this way, preparing a publication can lead to consequences far beyond the presentation of present results and can be a highly creative process.

But finally, you will have written a first draft that you are happy with. If you have co-authors, this is the time to discuss further improvement of the work with them, or to get their approval of the publication draft. Usually, at least minor corrections will then be suggested, and several iterations will be necessary until everyone is satisfied.

It is also helpful to ask more experienced senior colleagues who were not involved in the work at all to comment on it. Such internal reviewing and commenting undoubtedly helps to improve the quality of the publication. It helps you avoid harsh reviewer comments and achieve a greater likelihood of your article being accepted by a scientific journal.

3.2 Writing Style

3.2.1 Who Will Read Your Article and Where?

Before you start writing, you should consider the readership for which you intend to write. This should influence your writing style. be influenced by this. Also related to this is the question of how extensively to present background information and basics. Your article will fall on more fertile ground if it is written at such a level that the expected readers can understand it. Related to this is the question of the choice of journal. You should be aware that different journals have different writing styles. Specialized journals, which are aimed exclusively at professionals, have a more sober and scientifically or technically narrow style than more popular journalsthat are published for a wider readership, such as *Science* or *Nature*. In these journals, articles are also read by non-specialists who want to gain a broader overview. In order to be able to understand the results described in them at the forefront of science, the comprehensible presentation of background and fundamentals is particularly important. Also, the language used in them is more general and the terminology less subject-specific.

3.2.2 Write Clearly, Concisely and Logically

Regardless of the choice of journal, your style should be as clear as possible. If every second sentence can only be understood by reading in hard-to-find sources, then interest in your article will quickly wane, even among specialists. At best, people will immediately skip to the conclusions at the end of their publication to see if there might actually be anything interesting in the article hidden under the jumble of hard-to-follow information. For the same reason, jargonwhich reads like jargon, should be largely avoided. The same applies to the excessive use of largely unknown abbreviations.. It is essential that these are explained the first time they are used.

You should describe your work logically and compactly. Historical explanations of the field should be kept to a minimum in a journal article… be kept to a minimum in a journal article. Placing your work in a larger framework should suffice for this purpose. The style of exposition should also be didactic, which is especially important for articles in books. This includes comprehensible explanations and thoughtfully guiding the reader so that he or she is not left to comprehend complicated explanations alone. Negative examples of this are mathematics textbooks where in the middle of a complicated derivation it is written "as can be easily shown" – and it ends up taking the reader a long time to comprehend what is easy for the mathematics professor to show.

3.2.3 Scientific Expression

In scientific articles, the writing style is simpler and more formal than in books, newspaper articles or everyday language. In a sense, this style has developed as a by-product of scientific publishing. Just like the scientific style of work scientific writing must be logical, clear, well-structured and unambiguous. The vocabulary used is not extensive because scientific terminology is well defined and rather limited. Einstein once said that he got by in scientific English with only about 300 words. Flowery language, typical of fiction, is not consistent with scientific style. If you are not writing in your native language, it is advisable to consult dictionaries, including technical dictionaries, and the advice of colleagues with a better knowledge of foreign languages when in doubt. Familiarity with the scientific literature in your field will help you to choose the right technical terms and phrases. and phrasing. It is also useful for a beginner to note down essential phrases from other good publications for possible future use. However, do not copy complete sentences. Do not read outstanding academic articles written by native speakers of English from the point of view of content alone. Also try to learn something about the writing style and the way of expression. Be aware that you can become a better writer by becoming a more attentive reader of other publications.

Once you have written your draft written, you should discuss it with colleagues before submission, both in terms of content and presentation style. It is better to receive criticism from well-meaning colleagues than from disgruntled reviewers. You should ensure that your good results are presented appropriately and not slated in a poorly written article for secondary reasons.

If you are not absolutely fluent in the foreign language, it is safer to express yourself more simply. Concentrate on the essentials and avoid discussions of detail that lead offside. The help of colleagues with excellent English is useful for language improvement and adoption of the article.

In general, an impersonal style should be preferred. In multi-author publications, one can also say "we", "I" should be avoided. Instead of "I found" it is better to say "we find" or "one finds". The latter phrase implies reproducibility of results regardless of who is doing the research. Purists would tend to prefer a more impersonal style: 'it is found that …'. Such wording removes any individual involvement. However, in my opinion, this is a bit of an overstatement. You should use the passive form where it is appropriate, but not overdo it.

3.2.4 Importance of Good English

If English is not your native language, but you would like to publish your article in an international English-language journal you will face another challenge. A minimum requirement is comprehensible and largely error-free English, even if you do not exhaust the diversity of the language.

In our experience and that of many reviewer friends in editing articles, the most serious language problems occur with Chinese, Russian and Japanese authors. Although it often appears from the illustrations that the scientific results may be good, it is not uncommon for such articles to be rejected because the reviewer simply does not understand what the author is trying to say due to poor language. Therefore, it is highly recommended that all non-native English speakers have their articles reviewed by colleagues with better English skills before submission. More perfect English in the article increases the probability of acceptance.

As a non-native English speaker, what can you do to write better articles besides taking a language class?

- English scientific sentences should be short and usually no more than 15–20 words.
- Avoid claused language with subordinate clause within subordinate clause, which is typical of the German language.
- A sentence should contain only one thought.
- Use an active style. Germans tend to nounize verbs, which makes the language seem less active.
- Try to avoid the national typical mistakes, e.g. "false friends", i.e. words that sound the same in English as in German but have a different meaning, e.g. "concise" is not "pregnant" in English. Japanese people often confuse "r" and "l" because they are the same in Japanese. For example, when reading a conference announcement in a Japanese English-language magazine, I read "Blatisrava" instead of "Bratislava".

Ultimately, developing a good English writing style is a matter of practice. To speed up this learning process, it helps to consciously read many good publications by native speakers from a stylistic point of view. If you want to avoid frustrating reviewer comments, it is very useful to seek the advice of more experienced colleagues, at least for your first own publications.

3.3   Structure of Scientific Articles

For reporting scientific results in scientific publications, the following style has become widely established: The text starts with an "Introduction", continues with "Methods" or "Experiment", before the main part "Results and Discussion" – or divided into two parts as "Results" and "Discussion" – follows, and ends with a "Summary" or "Conclusions". This simple structure has become the basic structure of scientific articles or reports during the last century and is also related to the hierarchical structure of thinking in Western countries.

Western Style

The Western style of writing can be compared to a tree. It is a hierarchical and logical sequence of explanations, with each subsequent explanation building on the previous one. Nothing is discussed that has not been explained before.

Like a tree trunk, there is a broad general introduction at the beginning. The subsequent discussion can develop in different directions and become very ramified, like the branches of a tree. The top of the tree at the end are the conclusions, in which the various branches of the discussion are finally brought together again (Fig. 3.4).

Traditional Japanese Style

As a professor at Kyoto University, I was often asked by colleagues to review and rewrite their draft publications. In the process, it became apparent that the Japanese way of thinking is completely different from the Western way. For the Japanese, details are much more important, as reflected in Japanese garden art, gift wrapping, or the artful serving of food. Traditionally, Japanese thinking is not as hierarchically structured as in Western countries. Japanese first gather all the information and then form their picture, regardless of the order in which the information came in. The fact that the order in which information is plays a subordinate role in an article was often reflected in a sequence of discussion that was not logical to Western thinking. Occasionally, details that were still unexplained were started or conclusions were presented before the presentation of results. And then subsequent discussion explained

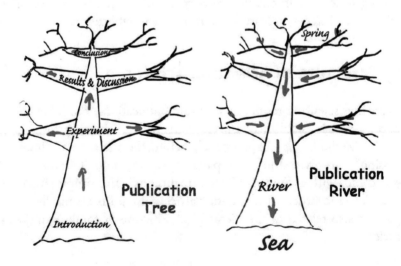

Fig. 3.4 Publication tree and publication flow

the initial assertion. Therefore, it can seem more like the formation of a river, formed from many small tributaries, in reverse to the structure of the tree, even if at first glance it looks the same from a distance.

Recommended Structure and Style

In the following, the structure mentioned at the beginning will be discussed in more detail. Compared to the basic four- or five-member structure that exists in principle, the current structure can be subdivided even more finely. The international acceptance of the basic structure has the advantage that scientists are familiar with it and can thus communicate with each other in a recognized structured way, in a "common language" so to speak. With English as the generally accepted language of science and a generally accepted style and publication structure, Heisenberg's vision of "science as a means of creating understanding between people" becomes somewhat truer.

Let us now take a closer look at the elements of a scientific article. For some articles, not all of the following statements apply, and in theoretical papers "experiment" is replaced by "theoretical methods". But most of what follows applies to all types of scientific publications, regardless of discipline.

3.3.1　Title

The title should be short, concise and informative. The goal is to catch the potential reader's attention. How do most readers of scientific journals, and probably you, go about checking the contents? The table of contents is read. If a title seems interesting, the abstract is read. Maybe the conclusions are read as well. Only then, when all this seems interesting and promises to be useful for one's own work, is the complete text read.

If a title is not worded in an interesting way and is too long, the article will have a harder time catching the attention of the journal reader. Therefore, you should formulate the title of your article as short and appealing as possible and in no case as a short form of the abstract. abstract. If you cover several topics in the article, focus only on the main topic and not on less important facets. If possible, the title should be no longer than one line and contain topical, attention-grabbing keywords. Superfluous phrases such as "A study by …" or "… measured with …" do not make the title more interesting.

Normally, you have to try to find a compromise between information content and brevity of the title. The final title is determined only after the full formulation of the article. Before completion, a working title is sufficient.

But do not underestimate the importance of choosing a good title. An illustrative example is Klaus von Klitzing's publication on the discovery of the quantum Hall effect. He had sent the paper to the journal *Physical Review Letters*. But it was rejected because the chosen title did not seem interesting enough to the reviewers and they did not immediately recognize the significance of the discovery presented under it. Fortunately, von Klitzing received a confidential tip that the article could still be accepted for publication with a better title. With a new title, it finally worked out in the second attempt. If von Klitzing had not been deeply convinced of the importance of his new discovery and had not done everything possible to publish this work after all, then a wrongly chosen title could almost have prevented him from winning the Nobel Prize.

3.3.2   Authors and Addresses

After the title of the publication, the author names are given on the next line, usually abbreviating the first names. It is customary for the first named author to be the principal author and, upon request, the corresponding author of the paper. Or a subsequent author is given as corresponding author. The names are followed by the addresses. If not all authors have the same address, the distinction can be made by superscript asterisks or numbers by the name.

3.3.3   Abstract

The abstract is a brief overview of the content of the article. It should not be more than about five lines long, so that it is not overlooked by readers in a hurry. It should describe the content of the paper, what was investigated with what, and the main result. It should by no means contain all conclusions or a complete summary of the article should be offered as an abstract. Also, a detailed discussion or citations are out of place here.

Like the title the abstract also has an informative and interest arousing function. The reader should feel motivated to want to read more of the article. Therefore, the abstract must be very thoroughly formulated and clearly present the essential aspects of the publication to the reader. Just like the title, the abstract should be formulated only at the end, when you know how you have presented everything.

Be aware that the title and abstract should be considered as one unit of information. as a single unit of information. This means that you should under no circumstances repeat the content of the title again in the abstract. By the way, this is a typical mistake of inexperienced authors, which is punished by attentive reviewers.

3.3.4 Key Words and Classification Codes

Many scientific journals give some keywords and classification codes after the abstract. *(keywords)* and classification codes for the PACS system (Physics and Astronomy Classification Codes) system. The PACS system, which was introduced and developed by the American Institute of Physics, represents a fine subdivision of the research fields. One can find here for a subfield the overview of the articles and books published in it. Similar systems exist for other disciplines.

Keywords are very useful for finding articles in an electronic search on the web.. If your article contains enough typical keywords, it may be found and cited. Forgetting to add the keywords will reduce the citation probability of your article.

3.3.5 Introduction

In the introduction must clearly state what the article in question is about. It should describe whether the work is experimentally or theoretically oriented, what the topics are, and what the whole thing is actually for. The motivation for the research carried out and its wider usefulness must be made clear.

Also, the existing state of knowledge should be outlined as a starting point for your work based on a thorough discussion of the essential articles published by others. The need for the submitted work should then be derived from the unanswered questions you have presented, with the aim of the current article clearly stated. But the attempt to sell one's results in the introduction as effectively as possible should not go so far as to raise expectations that cannot be met by the results in the end.

If the article is longer and consists of many parts, a few words about the structure of the work are appropriate.

3.3.6   Experiment and Method or Theory

This part of the article must provide the reader with all the information needed to understand, follow or check the content of the article. For experimentally oriented work, this includes sample preparation and analytical methods used, including details of the equipment used.

If you apply a self-developed method that is not known, something more must be said about it. Also, the evaluation of the data should be explained.

In the case of theoretically oriented publications, the theoretical or methodological bases and calculation procedures should be elaborated analogously, or the computer programs used should be specified. In the case of sociological studies, for example, something should be said here about the statistical totality considered and the statistical evaluation procedures, and in the case of legal studies about the legal basis.

3.3.7   Results and Discussion

This section is the main part of a scientific publication. In many science-oriented journals, the discussion of the results can be found immediately after they have been presented. In some journals, the presentation of results and discussion are divided into two parts. There are pros and cons for both types. On the one hand, it is more convenient for the reader to see the results presented discussed immediately rather than having to scroll back to the figures during the discussion. On the other hand, this style of presentation can be an amalgamation of valid, measured results and interpretations to be revised later on to be revised later. If there are many results to be presented before they are discussed comparatively, it is recommended that they be organized in two separate parts. Already when writing, orient yourself to the style of the journal to which you intend to send your article.

For the presentation of results figures and tables play a decisive role. Here, a graphical presentation is preferable to a tabular one, as the tendencies become more obvious in illustrations. However, do not rely on the fact that the reader can easily see from the graphical or tabular presentation of the results what he is supposed to see, even if everything is unambiguous for you. Always describe clearly in words what is shown in the figures and tables. and tables and what the relationships are.

When discussing results, it is better to develop a complex rather than a one-dimensional view. This can be done, among other things, by using different experimental methods.

You must not present your results in isolation in a vacuum. It is essential to establish a link to what has been published by other colleagues. In this way, you will be able to place the new knowledge you have produced appropriately in the existing state of knowledge. existing knowledge. Constructive discussion of the publications of other colleagues helps to arrive at a more comprehensive interpretation and to clarify any contradictions that may still exist. Read the publications in your field thoroughly. This knowledge will help you assess whether your work adds significant knowledge. Of course, in order to properly compare your results with the work of other colleagues, you should have understood the other articles and not misinterpreted them. Other scientists will be happy to see their interpretations quoted and confirmed in your article.

But what about articles that arrive at a different view from yours? In this case, these publications should at least be mentioned, because disagreement, apparent contradictions and unanswered questions often provide the impetus for further investigation, which later leads to full clarification.

Before submitting a publication, most scientists search for possibly overlooked articles with appropriate keywords in search engines or databases. You should also remember to do this before submitting your article in order to be able to give the most significant literature discussion possible as possible.

In the discussion, you should clearly elaborate what in your papers are new results or hypotheses and what constitutes a report on top of existing knowledge. If the two are mixed unmarked, you may be denied due credit because the newness of your article is not recognized.

Speculations and working hypotheses are often the starting point for new research. If you want to include them in your article, however, stick to more moderate speculationsIf you want to include them in your article, however, stick to more moderate speculations based on thorough stock and trend analysis, and don't get carried away with unsubstantiated and fantastic speculations. Niels Bohr once said about this: "It is difficult to make predictions, especially when they concern the future."

3.3.8 Conclusions or Summary

After you have presented and discussed your findings, the article should conclude with conclusions. For many authors, you will also find a summary of the main content at the end of the publication. This is useful for the reader who wants to get an overview of the article in a short form. It is usually written in the past tense, as it deals with what has already been laid out. But the summary usually does not contain any new information and is often a simple repetition of the key phrases of the discussion.

However, the better option is to instead present conclusions at the end of the article. These should attempt to arrive at a certain generalisation of the results. This is about highlighting the more general significance of the findings and placing them in a wider context. However, do not stray too far from the specific content of your publication. Formulate the conclusions as concisely as possible and do not dilute them with elements that are really only summaries. It is advisable to set the conclusions off against each other in sub-headings with touches and not to formulate them as a continuous text. This may also include comments on outstanding issues and suggestions for future work. When stating generally valid facts, the present tense should be used here.

Just like the abstract the conclusions must be particularly thorough and concise, as these are the two parts of an article that are read first and most often. If the conclusions are appealing, then the interest to read the whole article grows and the reader remembers the whole article better later.

A common mistake is to misleadingly call the summary Conclusions without including any conclusions other than restating what has already been said. Please take note of this.

3.3.9 Recognition

At the end of the article should not forget to thank all who participated in the success of the work. This refers to the colleagues and students whose contributions were included in the paper but were not sufficient, e.g. technical support, discussion of the results and interpretation of the publication draft. This is usually also where you mention the boss who created the framework or at least allowed the research to be carried out. Do not annoy these colleagues by tacitly passing over their contributions.

It is also necessary to explicitly name the sponsor of the project and to indicate the project number.

However, the experienced sceptic can sometimes read between the lines in acknowledgements, especially if he knows the people. If one reads in the acknowledgement, "I would like to thank Ms. P. Müller for technical support and Mr. G. Schneider for valuable discussions," then this can also mean, "Ms. Müller did all the experiments and Mr. Schneider interpreted them. But I am the boss and sell the results under my name."

3.3.10 Annexes

For lengthy calculations, additional data, or other space-consuming material that interrupts the flow of thought and does not necessarily need to be explained in detail at that point, it may be better to include it in an appendix rather than in the continuing discussion. But be careful with appendices. They certainly do not belong in a *short note,* which must be brief and focused. In such a case, where information is readily available elsewhere and the solution of a standard integral is lengthy, a reference to where the information can be found should suffice.

3.3.11 Quotations

The bibliography is an important part of any publication. The citations provide the link between your work and the pre-existing body of knowledge. Without such a classification, your results stand in a vacuum. After all, it is rather the exception that someone has realized a completely new idea where there was no preliminary work by other scientists.

For your publication to be taken seriously, it must relate to the most important other articles in the field. This is necessary not only to properly categorize your results, but also to show the reader that you have in-depth knowledge in your field. Also, citations are a matter of fairness. After all, by doing so, you are giving due credit to the colleagues who laid the groundwork and on whose work you are building.

3.3.12 Other Styles of *Short Notes* and *Letters*

The elements described above provide a useful structure for original work, and this is particularly true for experimentally oriented publications. Although neither structure nor length determines the quality of an article, you will avoid the risk of rejection on formal grounds if you follow the advice given. However, there are no hard incontrovertible rules for structure. The structure of *short notes* and *letters* is simpler, even if it is basically as stated above. Only the subheadings are omitted, and the text is much shorter and more concentrated. Above all, the reproduction of the experiment and the results as well as the discussion of results and literature are much less extensive than in an original paper.

An example of an outstanding *short note* is the article by Leo Esaki. This short article, which led to the Nobel Prize demonstrates conspicuously that it is not the length but the scientific content of an article that makes it valuable. The articles by Shockley, Brattain and Bardeen on the invention of the semiconductor transistor or by Hall on the Hall effect are equally short. What can you learn for your own work from such one-and-a-half page articles that led to the Nobel Prize? The conclusion is: You just have to have a good idea, perform a successful experiment, write a short article … and wait for the Nobel Prize.

3.4   Formal Aspects of Manuscript Preparation

There is a lot more to writing a scientific article than simply writing the text. In this section, we will look at the formal aspects of a scientific publication.

3.4.1   Figures and Tables

It goes without saying that the illustrations must be drawn thoroughly and cleanly. The correct parameters must correlate with each other, and the appropriate method of representation must be chosen, either linear, logarithmic or semi-logarithmic. Should it be a curve, a bar chart, a pie chart, or even a three-dimensional plot? The text in the figure and on the axes must still appear in at least 9 point font size when printed, so that everything remains legible. Also, the figure size should be appropriate: The illustrations must neither appear in postage stamp format nor too expansive.

For aesthetic reasons, all illustrations in a publication should be produced in the same style, i.e. the same frame and line thickness, the same fonts, and the same typeface.same fonts and font and size, etc. By producing all illustrations with the same program, you can certainly achieve this easily. Avoid lines that are too thin or give the impression of a funeral announcement in the frame thickness. Lines less than 0.2 mm thick may still be discernible on the computer screen, but can hardly be seen in print. Similarly, bold print in axis designations or figure texts should be avoided. This only distracts from the content. You will see for yourself whether your illustrations meet your aesthetic requirements or whether serious changes still need to be made.

As mentioned in Sect. 3.1.5, it is often useful to have a key mapping. This figure needs to be prepared particularly thoroughly. If your results are very significant, this figure may be reproduced citationally in other publications and still be recognized by later generations of scientists.

The rules for the design of the illustration are described in detail in Sect. 3.1.5 under point III. There you will also find a sample illustration recommended in the style as a template.

Each figure shall be accompanied by a legend. This text must be informative and brief and, above all, explain the axes, curves and symbols but must not contain a discussion of the figure or a description of the view to be derived. For the text accompanying the figure, the font size should normally be 9 point should be chosen. The figures, tables and formulas are to be numbered consecutively. Orientate yourself on the numbering style of the journal to which you intend to submit your article.

If you become the author of a chapter of an edited book, the chapter number is additionally indicated before the current number of the figure, table or formula, separated by a dot.

Tables must also be clearly laid out. For each column and row, it must be clearly stated what is being reported. The units must be in round brackets after the corresponding terms. As with the figure text, the font size should not be less than 9 points.

All figures and tables are also to be mentioned in the text. They must be placed as close as possible to the first mention, but must not be shown before the mention or even in a preceding section.

3.4.2   Units

In scientific and technical publications, the legal units of the SI system are generally to be used. are to be used. But sometimes there are exceptions to this. In such cases, use the units commonly used in your field. Avoid the use of obsolete or little-used units, because this makes your work more difficult to understand and less comprehensible. Also, in such a case, rewriting will probably be required by the reviewers. The basic units of the established Système International (SI) are s, kg, m, A, K, Cd and mol. In addition, there is a set of derived units endorsed by the International Union of Pure and Applied Physics (IUPAP), which include Hz, N, V, W, J, F, and Wb. Note that in scientific publications, including their electronic versions, the units are to be written upright and not slanted (italicized).

3.4.3   Abbreviations

The use of abbreviations and acronyms is very established in scientific texts and useful for saving space. But be aware that not every reader is familiar with all the abbreviations used. Often, someone who deals with these terms all the time tends to consider them common knowledge, while non-specialists have problems with them. Therefore, all abbreviations and acronyms used should be spelled out, defined, and named in full the first time they are mentioned, with the abbreviation or acronym must be given in brackets after it, for example "Reflection high energy electron diffraction" (RHEED). After that, only the abbreviations and acronyms can be used. Only if abbreviations and acronyms are widely known and accepted internationally, e.g. ppm, can they be used without further explanation.

Especially for books and also for longer articles, instead of explaining abbreviations when first used, it is useful to include a list of abbreviations and a list of symbols in the form of a list.

3.4.4   Symbols

> The symbols are so illuminating that the fact that the text is incomprehensible does not bother at all (A.N. Prior).

Just like acronyms and abbreviations symbols must also be explained when they are first mentioned; unless they are common knowledge.

Use only the established symbols unless you occasionally need to deviate from them for some reason.

For example, it is problematic to use "e" for exponent and electron charge at the same time in an article. Unless a different symbol is to be introduced for one of the two quantities, a way out in this situation is to write "e" once straight and the other time slanted as "*e*". Then the distinction would be given. It is also possible to add a footnote to avoid confusion.

3.4.5   Index

An index is only required for complete books. It makes it easier for the reader to find the desired information in the text. A well-structured index is not only alphabetical, but also contains the derived and secondary terms under the main terms. The generation of the index is made possible by modern word processing programmessuch as LaTex, makes it very easy. Here, the keywords are to be marked in the text, and then the index can be generated automatically.

3.4.6 The Bibliography

Importance of the Bibliography
The bibliography is an important part of any scientific publication and serves several purposes:

- It leads the reader to supporting, further, basic and detailed explanatory literature, for which books are particularly useful. In journal articles, experience has shown that it mainly concerns other articles, highlighting review articles that deal with the general overview, background, methods to the specific problem and also the measurement procedures.
- It recognizes the fundamental achievement of the scientists who laid the foundations of the field and essentially developed it. You are hereby stating on whose shoulders you stand. The work of these colleagues is usually acknowledged at the beginning of the introduction.
- The citation of the essential articles shows the reader that you have a good overview of your field and are able to evaluate your results. If you show your competence by doing this, the entire article will be more trusted.
- Conflicts of previous contradictory publications can be resolved by providing the key.
- You prove the correctness of your interpretation by citing similar and supporting results, and many authors like to prove their authority by preferentially citing the work of the most distinguished colleagues in the field.

How Many References Are Needed?
The number of literature sources depends strongly on the field of research and the type of article. While original articles in mathematics or engineering often have only about ten references, comparable publications in physics or chemistry have about 50 or 40 sources, respectively, and original biomedical works typically have about 100 references. The subject-specific differences are strongly related to the citation habits and research activities in the respective fields. and research activities in the fields concerned – and thus with the available and thus citable number of publications.

In *letters* or *short notes* there are significantly fewer literature sources, only about 20. A *letter to Nature, for* example, must not contain more than 15 references. In contrast, for a review article is required to contain well over 200 references, as it is intended to represent an entire field of research with all major publications.

Collect Literature Sources

It is no problem to select the papers to be cited if you have continuously followed the scientific literature and collected all the papers related to your research topic. and have collected all the papers related to your research topic. Being constantly informed is helpful not only for easily selecting the works to cite from your rich repository, but also for always receiving new ideas for your own research and interpretations from the professional literature.

I would like to recommend that you regularly collect the works that interest you, either in electronic form on your computer or as printouts/copies. I know from experience that this collection must be done in an orderly fashion, in thematically subdivided directories on the computer or printed out in binders and folders. Even if this means some extra work at the beginning, you can save much more time than spent later, if you don't have to rummage through high piles of literature or open hundreds of files. or having to open hundreds of files. It especially takes a lot of time if all you know is that there was an article by someone in some journal recently that you need to cite, but you haven't collected a copy of it.

Since most articles are now available online, they or the links to them can be easily to them can easily be saved to your computer. You will find that your literature collection grows rapidly as your work progresses. Therefore, a well-organized filing system, either physical or electronic, is a must.

Follow the Scientific Literature

It is much more efficient to constantly follow the scientific literature and to collect the relevant publications than to start searching for related literature only when writing an article. than to start looking for related literature when you're writing an article. This can then become incredibly time-consuming. For this reason, I would recommend that you set aside a few hours in your weekly schedule to read scholarly journals pertaining to your research area. Spending about 2 h a week reading the journals on your computer or in the library will keep you continuously informed of developments in your field. Otherwise, there is a risk that this task will keep getting put on the back burner under the pressure of the day's tasks. This is time well spent, even if it doesn't help solve the problems at hand at the moment. Nowadays, reviewing journals is much easier and faster than it was until the mid-1990s, as almost all international scientific journals are available electronically.

Secondary Sources

It is better to have actually read all the cited articles yourself and not to rely on taking literature sources from other publications. Blind copying can lead to faulty citation and misinterpretation. Copying erroneous citations convicts the bungler. As a general rule, referencing a paper can at least make it seem worth reading. If an article is not directly accessible to you, online or in print, the journal in question can also be ordered through interlibrary loan in most cases, or one can contact the author and ask for the electronic version or an offprint. Since most journals exist electronically, articles from journals not subscribed to at the Institute can also be obtained directly from the journal via *pay per view* get. Do not pass over articles that may be essential simply because you cannot easily obtain them.

Format of the Bibliography

Formatting the bibliography of a scientific publication can almost be considered an art in itself. There are several established citation styles, described in detail in [34]:

- In most scientific journals, references are numbered in the order of their use and appear in square brackets, e.g. [21]. In edited multi-author books, the chapter number is placed before the current number of appearance, e.g. [5.22]. In the bibliography, the reproduction is then ordered accordingly. For colleagues, who are not able to use the power of their word processing programs for the subsequent inclusion of literature sourcesFor colleagues who are not able to use their word processing programs, e.g. Latex or Microsoft Word, to their full capacity for automatic renumbering, the subsequent insertion of references, which leads to a change in the numbering, can become a problem.

In some older style journals, the source is the author's name and year of the article are given in round brackets, e.g. (Grunert 2004). At the end of the publication, the source references are then not reproduced in the order of use, but in the alphabetical order of names and, in the case of several literature sources by one author, in the order of the year below.

- Especially in humanities journals, but also in some biology journals, it is common practice to insert literature references as footnotes on the current page, whereby the citation number 37 is superscripted.

Even within these broad styles, different publishers do not have consistent formatting in the reference list. For example, in numbered references the year is given differently, at the end or in the middle between the volume number and the page number:

[1] J. Bardeen and W.H. Brattain, Phys. Rev. **74**, 230 (1948).

or

[2] B. Meyer, Nucl. Inst. Meth. **B 28** (1992) 186

To help you choose the right citation style right from the start of writing the article, why not take a look at some articles from the journal you want to send your paper to beforehand.

While there may still be some differing details, generally speaking, numbered literature entries:

- The first names are abbreviated. If there are two first names, there is no space between the initials.
- Journal names are abbreviated, and there is an international abbreviation convention that you must follow. You can find this list, introduced by CalTech, at: https://www.library.caltech.edu/journal-title-abbreviations
- The journal volume number appears in bold type.
- The first page of the article – and in exceptional cases also the last page – are separated by commas or hyphens after the journal number.
- The year of publication is given at the end in round brackets.

With alphabetical style, the source citation may appear in the reference list as:

Bardeen, J., and Brattain, W.H., Phys. Rev. **74**, 230 (1948).

or

Schmal, G., Z. (1994), Z. Onkologie **39**, 34

Here the initials are placed after. The year of publication of the journal is either in the "normal" citation style at the end or appears after the name as in the text. Books or contributions from them and doctoral dissertations are cited differently again. In the numbering system a book would be cited as:

1 P. Yu, M. Cardona, Fundamentals of Semiconductors, 3rd edn., (Springer, Berlin, Heidelberg 2001), p. 377

or in the name-year system as

Yu, P., Cardona, M. (2001): Fundamentals of Semiconductors, 3rd edn., (Springer, Berlin, Heidelberg 2001), p. 377

Assuming your article is perfect in content and layout, it is not likely to be rejected for incorrect formatting of the bibliography. But you should always try to get as close as possible to the style of the desired journal.

Especially in publications with a large number of authors, should all authors always be listed in the literature entry? In such cases, one often sees only the first author listed, and the co-authors disappear under et al. disappear under et al. to save space. However, for articles with up to three authors, all authors must be listed in the citation. Detailed instructions on this can be found in the Chicago Manual Style [34] for detailed instructions.

Most publishers have their own rules for formatting bibliographies. which can be found either in the journal or on the corresponding webpage.

Also note that many online journals belong to the *cross-reference*-agreement (a system that allows users to go directly from the citation of an article to that article) and have therefore introduced special citation rules. These should also be followed where appropriate.

3.4.7 Camera-Ready Manuscripts

Many publishers require the submission of camera-ready manuscripts for academic publishing, both for journal articles and books. This is due to cost considerations and shorter publication deadlines. The same electronic data used to print the article or book is also used for the online version of the book or journal.

In order for authors to submit their manuscripts in the required layout, publishers provide Latex or Word macros available. There is no doubt that LaTeX is the better and more powerful word processor; however, it is more difficult to use than Word and takes some training to master. But if you're going to be publishing a lot in the future, learning LaTeX is well worth the time investment.

The main advantages of this typesetting program are:

- Any kind of numbering is done automatically (sections, figures, tables, equations, citations, etc.).
- It is best suited for the perfect reproduction of complicated mathematical formulas.
- It provides the best typography and text formatting. A manuscript created with LaTeX looks professional, while Word manuscripts often seem homemade.

- It is widely used within the mathematical and physical research community. Publishers prefer LaTeX because of its perfect layout and easier handling than Word, especially when including figures.

Once you have learned LaTeX, you will enjoy it and won't want to miss it.

3.4.8   Some Remarks on Copyright

With the advent of printing also arose the question of copyright. In the late fifteenth and sixteenth centuries, the problem of copyrights became increasingly precarious due to numerous pirated copies, so that in 1710 the British Parliament addressed this problem formally and legally for the first time in the Statute of Anne. Although initially primarily concerned with the rights of printers and letterpress printers, the scope was later extended to include authors' rights in copyright.

Today's Copyright Laws
The copyright holder has the right to display, reproduce, distribute and license the work in question. Conversely, it is an infringement of the copyright if other persons or organisations exploit a publication or parts of it in an unauthorised manner. For books still on sale, heirs hold copyright for up to 70 years after the death of the last surviving author. Exceptions to this exist in Japan and Canada, where a minimum period of 50 years applies, as set out in the 1971 Berne Convention on Copyright.

Although copyright was originally about publishing and authors' rights in writing, with the development of new technologies the term "writing" has taken on a more general interpretation. Copyright is now also about architectural design, graphic art, software, films and music. Something can only become the subject of copyright if it is original and expressed in a specific material medium.

Copyright law is quite complex. Especially in the context of the development of electronic media and the Internet, the rules for copyright protection have had to be constantly updated and expanded. Currently, the World Intellectual Property Organization, abbreviated as WIPO, is dealing with copyright requirements in the information age and making proposals for updating.

Copyright in Scientific Publishing

In the case of scientific articles and books, the author normally assigns the copyright to the publisher, who thereby assumes the right to print and distribute the work.who thereby assumes the right to print and distribute the work. The authors, on the other hand, benefit from the dissemination and popularization of their work and the additional service offered by the publishers.

If you are preparing your own publication, the first copyright question often concerns the use of material, e.g. illustrations, from other publications. If you want to use figures, tables or text parts from other articles, then you must obtain the copyright for them, even for material from your own publications. This applies not only to external material, but also to parts of your own previous publications, such as figures, tables and text parts. You must then ask the first publishing house for permission in order to be on the safe side legally. Because by signing the Copyright Transfer Agreement you have transferred all rights of use to your article to the publisher. This permission is almost always granted without any problems, especially when it comes to your own work. A condition for the adoption of parts of other publications is – no matter whether foreign or own quotation -is that the origin is correctly stated. This means that you have to indicate the citation at least in the figure text.

Some publishers charge a fee for the copyright grant.

3.5 Submission, Peer Review and Revision

Once your article is written, with a particular journal's style already in mind, and all co-authors are happy with it, it's time to submit. You can find the submission address in the journal.

3.5.1 Appraisal

Normally, you will receive an acknowledgement of receipt immediately after your publication has been received. First, the article goes to the journal's co-editor responsible for the field in question. He then sends the paper to usually two reviewers, if he does not review it himself. The independent reviewers are generally internationally recognized experts in the field.

If both reviewers agree, the article is accepted for publication. If the results differ, a third reviewer is consulted.

Do you have the possibility to influence the selection of reviewers for your article? Most journals do not consider reviewer suggestions from the authors because they assume that these are only good acquaintances of the authors who do not provide objective commentary. Journal editors select the reviewers themselves, recognized experts in the field. But you have the option to exclude potential reviewers. If you ask when you submit your publication not to send your paper to certain colleagues because they are competitors or you have had negative experiences with them, they will usually comply.

But sometimes a journal editor has no idea to whom to send the paper for review because perhaps the field is too new. Then they check the bibliography to see who might apparently be a competent colleague to do it because their publications are cited as foundational or multiple. In this way, you can occasionally influence the reviewer's selection of your literature. You can then expect to find a more favorable judge because everyone is happy to see their article cited in yours.

The review is anonymous because this is to avoid conflicts building up. But sometimes you can infer the reviewer if the commentary recommends citing additional works and an author's name appears more than once. This is because it can improve his own citation rate.

Regardless of how the reviewers are selected, *peer reviewing* is an indispensable prerequisite for maintaining the quality standards of scientific publishing. It is a cornerstone of quality control. If an article has been accepted by a journal, it meets at least basic quality standards.

But there are also a number of potential problems with the review process:

- Not every reviewer writes his assessment quickly enough, which often unnecessarily delays the publication of an article. In extreme cases, the reviewer has to be changed. The pressure on reviewers to produce an assessment quickly is growing, especially in connection with electronic and online-first publishing.because the aim is to drastically shorten the time between submission and publication of an article, from an average of 6 months for printed publications to about 1 month.
- Not every review is really an evaluation of the content. Occasionally the content is repeated rather without comment.
- Another question is about the competence of the reviewer. Therefore, he is asked by the journal that sends him an article for evaluation whether he feels competent. If not, he must return the article or pass it on to a more competent colleague. From the 14 publications by J. H. Schön in the journals *Science* and *Nature in* only 1 year, it can be seen that none of the reviewers realized that these results could not be true.
- In addition, an appraiser is asked whether he or she sees no conflict of interest. If he or she is not in a position to provide an objective review

because he considers the author of the work to be reviewed as a competitor, then he should refrain from the review. But occasionally, this is precisely what is seen as an opportunity to thwart someone and come out ahead yourself in the process. After all, a reviewer can easily delay the appearance of a publication by demanding unnecessary and time-consuming rework. A number of cases are known, especially for American journals, in which this artificial delay in publication resulted in the reviewer's group achieving the first and thus most cited publications in new fields, even if the submission date of the artificially delayed paper was older and thus the priorities were secured. Because Bednorz and Müller wanted to ensure that they really were the first to publish an article on high-temperature superconductivity, they did not choose an American journal for this purpose, but one published in Germany.

3.5.2 Acceptance or Rejection

At Based on the reviewer's comments, the journal editor will choose one of the following three options:

- The article is rejected.
- The article is accepted in principle, but some revision is requested.
- The article is accepted unconditionally as it is.

Naturally, the most prestigious journals have the highest rejection rate because they are offered many more articles than they can publish. For example, at the leading scientific journals *Nature* and *Science,* only about 5% of submitted articles are accepted. For the other leading international journals, this rate is around 30%. By separating the wheat from the chaff in this way, the quality of the journals is maintained.

But the rejection can also occur for reasons other than quality. For example, it may be judged that an article does not fully fit the profile of the journal, or that it is scientifically insufficient or does not contain enough new information. Or the writing style does not match that of the journal. Therefore, it is recommended that you have thoroughly familiarized yourself with the profile of the journal of choice and the style of other articles in it before submitting your article. For example, the style in *Nature* is more popular than in purely professional journals because it is intended to appeal to readers who want information beyond their own field. Also, before submitting an article, you should have checked how long comparable articles are in the journal of your choice so that it is not rejected for being too long.

You should not neglect the thorough selection of citations. If the level of knowledge in an article is not adequately reflected by the appropriate citations, or self-citations dominate, doubts arise about the author's overview.

The likelihood of acceptance increases if the English is good.

Also, acceptance of an article may be refused because expectations were raised in the headline and abstract that the subsequent article does not meet.

If you keep all of this in mind, then you can avoid wasting time on submissions and having a negative experience when your article is accepted.

3.5.3   Discussion with the Reviewers

Of course you are free to resubmit an unaccepted article to another journal so that all the work is not wasted. But always examine the reviewers' suggestions critically and improve your article largely as suggested, provided you accept the comments. Under no circumstances should you resubmit an article rejected by one journal unchanged to another journal. It is not uncommon for the new journal to appoint the same high-profile colleague as a reviewer. If the article still contains the old points of criticism, it will be rejected there as well. But if you have improved the article as recommended, then your chances increase.

But what should you do if the reviewer's commentary is unfair or shows a lack of understanding? You do not have to accept everything slavishly and can certainly respond to it with a corrective letter. Sometimes this discussion helps. But don't take the wrong tone by becoming aggressive. This will not help your cause at all. This will only anger reviewers and editors alike, and will not bring you any closer to resolving the conflict. If no agreement is reached, then withdrawing the article… and resubmitting to another journal is the only way out. However, it will not be credited as a clever idea for an author, in anticipation of possible submission problems, to in anticipation of possible submission problems to avoid time delays. Imagine how a reviewer might react to receiving the same article from different journals at the same time.

It is worth considering, however, whether one should have the forehead to resubmit an article when one receives the following comment, "This article contains much that is new and much that is true. But unfortunately, what is true is not new and what is new is not true."

3.5.4 Happy Ending

If everything went smoothly and your publication was successfully accepted, then congratulations. You have obviously taken into account the advice given in these pages.

After your paper is accepted, it will generally take another two to 12 months to appear in the journal, 6 months on average. Since all journals are also accessible via the Internet, your article can be made known more quickly.

3.6 Writing Book Manuscripts

I would now like to make a few brief remarks on a subject that is probably not relevant to you at present, but could become so in the future: book writing.. Being prepared in time helps to better cope with later requirements.

3.6.1 Writing a Contribution for an Edited Book

It may well happen that you are asked to present your outstanding scientific results in a chapter in an edited book. This honour is comparable to being asked to write a review article for a journal. After all, only the most respected representatives of the field of knowledge should be present in a book.

In a book contribution, the aim is first to present the basics of the field in question, then to give an evaluative overview of the relevant international state of knowledge, similar to a review article, and finally to present the latest developments including the open questions. Your own results should be appropriately integrated, but not necessarily dominate the foreground.

In order to make the book as coherent as possible, it is useful to agree on the content with the other contributing authors before starting to write. Cross-references to the other chapters also increase the consistency of the presentation.

In terms of writing style and breadth of subject matter and the breadth of the subject matter, a book contribution can certainly be compared with a review article. In contrast to this, however, the basics should also be explained in an introductory way for better understanding and there should be a coordination of content with the other book contributions, neither of which needs to be observed when writing an article.

3.6.2   Editing a Book

If you receive an invitation to edit a book, it also implies that you are recognized as a leader in your field and that your work is highly regarded. What are your tasks as an editor of a book?

- First, in consultation with the publisher, decide whether you will become the sole editor or bring in one or more competent colleagues as co-editors with whom you will share the work. It helps to have leading colleagues as co-editors in each subfield.
- The next step is to gain clarity about the planned content of the book, i.e. to draw up the table of contents. At the same time, the number of pages per contribution must be decided, whereby the weight of the content of the individual sub-topics should also be reflected to some extent in the proportion of pages.
- Once the content concept is in place, it is then necessary to determine who will write the chapters. For this purpose, contact must be made with the relevant co-authors. The presentation of the overall content concept and the significance of the expected contribution of the co-author can help to overcome initial reluctance towards the further and new challenge of writing a book contribution. Most scientists are more willing to write a chapter in a book than to write a whole book on their own.
- Now that the team of authors has been decided upon, each contributing author should write down the detailed content plan for his or her chapter and discuss it with the book editor. Once these plans have been fixed in the discussion, they should be distributed to all co-authors. This way, everyone knows what the other co-authors will write. In this way, repetitive discussions can be avoided and cross-references can be given, making the whole book more coherent.
- To ensure consistent terminology and symbolism, this should be agreed at the outset.
- The layout of the book will be more uniform if sample illustrations have been distributed by the editor at the beginning, and all contributing authors must follow this style.
- Uniformity of presentation can also be greatly enhanced by each co-author receiving a sample chapter from the editor as a template for writing style and following it.

3.6.3   Writing a Book Manuscript Yourself

It's a lot more work to write an entire book manuscript… on its own. A scientific book is expected to present the basics in an understandable way, to describe and explain the subject in depth, and to reach to the most advanced applications. It must remain valid much longer than an article.

3.6.4   Writing to Make Money?

For No honoraria are paid for articles in scientific journals, except for invited review articles. Most scientists consider publishing their results as a normal and important part of their work. Since they are aware that good publications are a prerequisite for achieving an excellent scientific reputation and can be very career-enhancing, no additional financial recognition is expected for publications. Moreover, most publications are written in their spare time because they often lack the peace and time for it in their strenuous day-to-day work.

Einstein, for example, worked on a number of his groundbreaking publications secretly alongside his completely different job at the patent office (he had a private box in his desk for this purpose, which he quickly closed when colleagues entered) or while he had his children on his lap in the evenings.

For books, the financial situation is somewhat better for the authors. Here, either a small one-off amount *(flat fee)* or a sales-dependent author's fee is paid for complete books. But only in a few exceptional cases does the author of a scientific book earn real money, such as Simon Sze, whose semiconductor textbook sold 1,000,000 copies. Only with standard textbooks and popular books about science does a science author possibly make an acceptable hourly wage. This is because science books written for smaller audiences do not usually achieve the kind of print runs or circulation that popular books, such as *Harry Potter,* do. One book author we know calculated his hourly wage at five cents when he applied the fee he received to the number of hours he invested.

Accordingly, the real benefit a scientist derives from scientific writing is not material. It is the satisfaction of having documented the knowledge one has accumulated over a long period of time and of having contributed a little to world knowledge. Of course, there is also a certain pride and the attempt to gain a reputation as a recognized expert and to "sell" the results as well as possible. This means publishing them in the most prestigious, best-cited and widely circulated journals, or choosing the book publishers with the best international distribution network. If you are convinced you have written an excellent article, send it to the top journals. Try your hardest, but still remain realistic. Remember that the printed word is worth no more than the paper it's written on until it's read by someone who will benefit from it.

Not everything printed represents the same value (Fig. 3.5): identical paper and matching size, but very different value.

The same applies to a specialist publication: its scientific content determines its value. But sometimes the paper on which a publication was printed still has some influence. A greater value is attached to an article that appears in the specialist journal *Nature* than to an article in a smaller national journal.

Fig. 3.5 Not everything printed on paper has the same value

3.7 What to Do and What Not to Do

Do	Select the appropriate content for your article.
Don't	Do not overload the article with irrelevant or peripheral information.
Do	Select the appropriate article type for your work.
Don't	Don't block the possibility of an outstanding, in-depth publication by hastily publishing a *short note*.
Do	Choose the right publication time.
Don't	Don't reduce the impact of your work by publishing too early or too late. Do not miss out on a possible patent application.
Do	Structure your publication logically.
Don't	Do not discuss results before they have been presented.
Do	Aim to publish in a reputable scientific journal.
Don't	Don't hide your results by publishing in a journal that is either low level, low circulation, or non-English speaking.
Do	Publish only true and confirmed results.
Don't	Don't sacrifice thoroughness for publication speed.
Do	List all colleagues as authors who made significant contributions to the article.
Don't	Don't forget to acknowledge any support from colleagues and funders.
Do	Follow the rules and guidelines of copyright.
Don't	Refrain from any kind of plagiarism.
Do	Discuss your findings with reference to what has been published by others.
Don't	Don't forget to cite the essential works.
Do	Choose an attractive and short title. Make sure that the abstract and conclusions are to the point.
Don't	Avoid cramming too much detailed information into the title, abstract, and conclusion.
Do	Follow the formal requirements of the journals in terms of writing style, length, formatting and language quality.
Don't	Don't be upset if you are asked to revise your article.

4

Culture and Ethics of Scientific Publishing

After the specific advice on writing your own scientific article in Chap. 3, I would like to make some more general remarks on the culture and ethics of scientific publishing and discuss some issues of form. Readers interested solely in article-writing advice may omit this chapter, at the risk of missing some interesting side issues.

4.1 Purpose of Scientific Publishing

In this section we will see that scholarly publishing serves several purposes. These can be summarized as follows:

- Documentation of knowledge,
- Discussion forum as an engine for further progress,
- Delivery of Information,
- Teaching,
- Establishment of priority rights,
- Support for scientific careers.

4.1.1 Obligation to Publish

It is the task of a scientist to create new knowledge, as the word "science" suggests: "to create knowledge". Many scientists turn to their field of research out of a certain scientific curiosity and to find intellectual satisfaction in

C. Ascheron, *Scientific publishing and presentation*, https://doi.org/10.1007/978-3-662-66404-9_4

discovering new things and contributing to the knowledge of mankind. But scientific research is not supported to allow scientists their intellectual pleasure alone. The goal is to contribute to the total knowledge, to improve the world through new insights.

If a scientist conducted his research without publishing the results, it would be like a company producing something but not selling it. And just as the company would go bankrupt, the scientist would lose the necessary support for his work if he did not publicize its results. Eventually, he would have to stop his research. Because most research support... comes directly or indirectly from public funds, it is a moral duty for a scientist not only to produce the new knowledge, but also to disseminate it.

Through publication in scientific journals new knowledge is both made available and stored for present and future generations of scientists. In this way, each individual publication makes a small contribution to world knowledge.

4.1.2   Driving Scientific Progress

Scientific publications represent an important forum for the exchange of information and discussion of scientific results. Before a consensus on new interpretations is reached, several publications by different authors on the same topic and an intensive discussion of the new results therein are necessary. To avoid duplication of work and thus counteract the waste of intellectual capacity and research potential, it is essential to continuously follow the scientific literature. If you read the scientific literature thoroughly before starting a new research project, you can avoid reinventing the wheel.

Without enough substantial publications, most fields of research could hardly have made progress. Just look at the development of X-ray physics or high-temperature superconductivity.. Fundamental papers [36] have triggered an avalanche of further publications. In Fig. 4.1, the influence of the 1986 article by Bednorz and Müller on high-temperature superconductivity [37] can be clearly seen.

4.1.3   Securing Priorities

Another important function of scientific publishing is to secure priorities on the initial discovery or initial development of ideas. Priority rights are

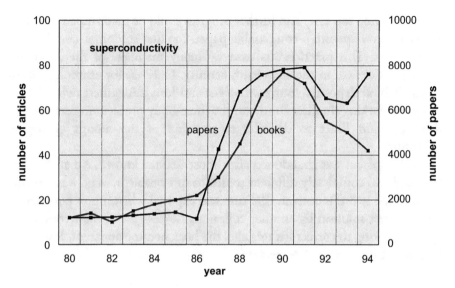

Fig. 4.1 Development of the number of published articles on superconductivity in the first years after the discovery of high-temperature superconductivity

determined by the submission date and not by the publication date of the work. Most scientists attach great importance to priority rights.

This is very understandable, because a documented outstanding scientific discovery can have a significant influence on the reputation and scientific career of a scientist. scientific career of a scientist, up to and including the Nobel Prize. Often, new effects are named after their discoverer, i.e. the person who first reported them in the literature, think for example of Planck's quantum of action, X-rays, the Haber process or Hodgkins syndrome, to name but a few. Nobel prizes are always due to the corresponding publication, as written documentation of the result is required.

In 2003, a few groups, one at the Max Planck Institute of Quantum Optics in Munich, one at MIT and one at the National Institute of Standards in Boulder were working on developing the atomic laser. It was a neck-and-neck race. And finally, the American colleagues and the MPG group submitted their publications on the subject, with only a week's difference. The slight delay was enough for the US groups to win the Nobel Prize.

As recently as the last century, there was a trick scientists could use to secure their priorities before publishing their results in summary form. It was permitted to send to a journal a sealed envelope, a so-called *pli cacheté, to* a journal. This documented the date of submission by the receipt of the mail. The author decided on the opening and publication date. In this way, competitors were not alerted too early and further progress could be made quietly. Now, if someone came up with a comparable paper, the first submitter could secure his priorities. This practice was used by many members of the Académie Royale des Sciences in the eighteenth century. In 1747, for example, Alexis-Claud Clairaut deposited four *plis cachetés* and Jean d'Alembert two, because they were working simultaneously on the three-body problem in competition with Leonhard Euler to describe the motion of the moons of Jupiter and Saturn.

But this system was abandoned after abuse became known. An author had deposited two completely different solutions to a problem, as he was not sure which of the two was correct. The ruse was exposed when, upon publication of the correct solution by another colleague, he designated the wrong envelope. By seeking to correct his mistake, the hoax became apparent, leading to the end of the practice of *pli cacheté*.

4.1.4   Further Benefits for the Individual

All active researchers regularly read the new scientific literatureespecially journals, to keep abreast of advances in their field. Even advanced students working on their thesis or doctorate, and postgraduate researchers who have to familiarize themselves with a new field, must first familiarize themselves with the relevant scientific literature in order to be up to date with the current state of knowledge.

Without the thorough reading of the scientific literature, many dissertations and publications would only represent the repetition of what has already been published and would not be accepted if the reviewers check thoroughly (the condition for acceptance is the new solution of a scientific problem). To avoid such a waste of time, the relevant journals must be read regularly, and use should also be made of electronic information systems.

The written documentation and availability of accumulated knowledge – whether printed or electronic – is an indispensable prerequisite for scientific progress. Before the invention of printing and writing, knowledge could only be passed down from generation to generation orally.

Even though our brain is very powerful, it cannot store nearly as much information as is documented in total in writing. If we know where to find what we are looking for, be it in a book, a magazine or in electronic information media, we can free our brain from memory work and have more capacity for more creative work. This aspect also supports the rapid progress of science in modern times.

Publishing one's own new ideas and results is essential for a scientist's recognition in the international scientific community and for career development.. If you receive frequent invitations to give keynote or plenary lectures at major international conferences, to write review articles, book chapters or books, then you have made it and can consider yourself well established. Until that is achieved, you need to pay close attention to your publishing activities. The list of articles published and papers given is, along with citation frequency a major criterion for success in being selected for a position.[1]

Scientific publications are also important when awarding research projects. The more extensive your list of publications is, the better your chances of reaching the next level of your professional development.

There are no fixed rules for how many publications one must have at which level, but some general empirical values can be given. It is not required to have a publication in conjunction with the thesis. to have. But a good thesis should lead to a publication of the results, usually as co-author with the supervisor.

In the course of writing a dissertation, some publications should be produced. These can be original papers, short publications in journals or conference presentations (documented by the conference proceedings). The required number varies greatly from university to university. In Japan one article was sufficient, at some German universities at least five articles are expected.

As a postgraduate assistant (or *assistant professor* at American universities), it is on the way to *tenure* (a system that exists at only a few universities in Germany, and the development from *assistant professor* to *associate professor* to *full professor* at the same university), it is an absolute must to write many good publications. Before entering the next level, the candidate's scientific activities are critically evaluated. In order to reach the professor level, whether in Europe, the USA or Asia, at least 20 original papers, 50 short publications *(short notes* or *letters)* and one review article are expected.

In this way, the milestones you have set in the course of your research are documented. Whereas in Germany the Habilitation was an indispensable prerequisite for an appointment to the professorship Nowadays, achievements comparable to a habilitation are also recognised, which can be an *associate professorship* at a foreign university or the management of a larger research unit in an industrial laboratory, if this is supported by corresponding publications.

---

[1] The condition for the recognition of citability is the publication of the article in a regular journal with an ISSN or a regular book with an ISBN. ISSN means International Standard Serial Number and is the identification code of periodicals. Every scientific journal and book series has an eight-digit ISSN assigned by the International Serials Data System in Paris. For books, the ISBN (International Standard Book Number) is assigned by the Library of Congress in Washington. This ten-digit number characterizes the language domain of the corresponding publisher, the publisher itself, and the individual book, and ends with a check digit to prevent confusion.

Writing a habilitation has also been made much easier by the fact that a coherent collection of publications in conjunction with an overarching preface is recognised as a habilitation thesis.

But one should not believe that the number of publications alone is decisive. Then someone who sells his results in small portions in a salami tactic would be in a better position than someone who writes few but more extensive publications. The quality of the papers is also assessed. An additional criterion that has come into use since about 1995 is the evaluation of the influence *(impact)* that a publication has, expressed in the frequency of citations in publications by other colleagues. This counting is done by the Science Citation Index which is created by the Institute for Scientific Information (ISI).

Frequently cited papers are rated higher and bring more prestige to the candidate than papers that are rarely or never cited. But this criterion should be applied appropriately: There is a much greater chance of getting highly cited papers in the mainstream of research in fields where many scientists are active than in smaller research fields or when a new research field is opened. This is why, for example, there are few Nobel Prize winners among the 100 most cited physicists. It is also wrong to rank publications only by the prestige or *impact factor* (average citation rate) of the journal in which the paper appeared, as is erroneously done by many appeal committees. The *impact factor of* a journal is not representative of the individual article. On the one hand, about 50% of all publications are never cited, and on the other hand, there are many articles that exceed the average citation rate of the corresponding journal.

Once someone reaches tenure… then the pressure to publish decreases somewhat. The positive side of this is that a professor then has more time for teaching tasks, project procurement and leading the research group. In publishing, the professor benefits from having undergraduate and graduate students in the group producing results and the leader becoming a co-author. The role of a professor as a mentor, trailblazer, driving force, informed discussion partnerideas provider, project fundraiser, etc. qualifies for co-authorship of the publications of students and younger colleagues. If a professor is successful in these tasks and continues to initiate excellent scientific work, then he has achieved the goal of many strenuous years and also enjoys recognition in the international circle of colleagues.

4.2   Types of Scientific Publications

Having already discussed the details of scientific publications, it is useful to give an overview of the different types of scientific publications. I will start with the simplest and quickest type and then move on to the more sophisticated, elaborate and conventional types of publication.

4.2.1   Own Homepage or Preprint Server

The easiest and fastest way to publish your discoveries, i.e. "make them public", is via your own or the institute's homepage.

However, if you try to make your results known in this way alone, this has the disadvantage that only a very limited number of scientists will become aware of them. In addition, this type of publication is not subject to any kind of quality control by *peer* reviewers *(peer reviewing), which* is why such publications are accorded only minor importance. These "pseudo-publications" are recommended as an additional, but not the only option. Here it is difficult to secure priority rights as with a journal article.

Place your work on a well-organized and recognized preprint server.the likelihood that your work will be read and possibly cited is higher. It can also be cited with the Digital Object Identifier (DOI), just like a journal article. Currently, the most widely accepted preprint server in the field of physics is the e-print archive http//:arxiv.org . This preprint server is organized as well as a scientific journal and is used by many scientists. There is also a certain initial hurdle for being able to place an article there: One must have a previous publication activity in journals or needs a recommendation from a respected scientist.

There is no quality control and confirmation of the results by experts in advance as in scientific journals *(peer reviewing),* but there are comments afterwards. Only those articles that later appear in scientific journals meet this criterion. Journals often allow articles to be placed on a preprint server before they are "properly" published.

4.2.2   Conference Proceedings

The material presented at conferences in the form of lectures or posters is subsequently documented in conference *proceedings,* documented. The conference proceedings can be a special issue of a regular journal or a book.

If you have a conference paper in a journal or regular book with an ISBN, that is the easiest way to get a citable publication. Some conferences only publish abstract volumes. However, these are not citable publications.

Normally, conference proceedings have strict size limits, e.g. two pages for short papers, four pages for invited keynote presentations, six pages for plenary presentations. Because the results cannot be discussed comprehensively in these, this is the only type of publication where republication of the same material in a different medium is acceptable, but only if the content goes substantially further. Otherwise, results may only be published once.

4.2.3 Publications in Journals

The most important and effective channel for the rapid dissemination of newly acquired knowledge is journal publications.

Types of Journal Articles
Journal articles are divided into three main categories:

Review Article

These are relatively long and comprehensive articles that provide a summary status report on the state of research in a specific field. They are usually written only by leading experts at the invitation of a journal. Many journals have one or two review articles at the beginning. However, there are also journals that publish only review articles, such as *Review of Modern Physics*. The usual length of review articles is between 20 and 100 pages, and in exceptional cases up to 200 pages, so they can almost reach book size. Unlike an original paper, the main content of a review article need not result from the author's own work. This is because it provides a critical overview of the state of knowledge in a specific field, i.e. it essentially discusses the work of other colleagues. However, the author's own work often takes up broader space in a review article. A good review article should present the field as comprehensively as possible, starting with the pioneering work in the field and including the most recent significant publications. It is particularly useful for newcomers to get an overview of the new research area from a review article, in addition to reading the relevant books. Reading a review article is also very useful for students in the qualifying phase. The purpose of a review article, similar to that of a book, is to present a field comprehensively. Even with very long review articles, however, there is still a significant difference from a book. In journal articles, due to space limitations, the basics are not laid out. Because of their comprehensive content and consequent broader appeal, review articles are cited much more frequently than any other articles. But often citations of review articles are actually borrowed citations of the other articles discussed in them, because readers are simply too lazy to read a variety of other original papers.

Original Works

This type of article serves as a comprehensive discussion of original results by active scientists. In it, significant new results of their own, of an experimental or theoretical nature, the manner in which they were obtained, and their interpretation and implications are presented. Such an article is usually written when a research project is nearly complete and the author has achieved such a comprehensive understanding as to develop a complex picture. The new results must be appropriately classified through extensive literature discussion appropriately, discussing not only supportive but also contradictory findings of other colleagues. Citations are almost as important as the new content in such an article. The length of original papers varies greatly around the typical page limit of about ten journal pages.

Short Publications (*Letters* or *Short Notes*)

When the aim is to publish significant new results quickly and to secure priority rights, shorter publications that can be written more quickly are chosen. These short articles are limited to about one to five pages, depending on the journal chosen. They also serve to generate broader interest in a new area of research. Often such papers are published as a simple presentation of new results before the full interpretation is available. After further studies have been conducted, a more fully discussing original paper may then follow. To avoid this being seen as republishing this article should contain at least about 50% additional results. Also, parts of the previous publication must not be repeated verbatim.

Since the international journals are published less in paper form, but essentially electronically and are accessible via the Internet, the articles can reach the interested reader much more quickly. This is especially true for the natural sciences.

Leading Magazines

First of all, we should ask the question when a journal can be called "leading". The main criterion is the ranking number, which is determined by the *impact factor.* which is determined by the impact factor. The *impact factor* characterizes the average number of citations per article of a journal within the second and third year after publication and is calculated by the Institute for Scientific Information (ISI).[2] In this way, the *impact factor* also reflects how many scientists read a journal and how much attention the articles published there receive, which is commonly equated with the importance of these articles. However, as with all statistics, this one should not be over-interpreted. It provides additional useful statistical information, but should not be used as the sole criterion for evaluating the quality of a journal. One has to be aware that the *impact factor* of a journal is essentially determined by the 10% more highly cited articles, while the remaining 90% have less influence on it. For example, the *impact factor of* the leading journal *Nature* increased by 20% in 2003 by just two extremely highly cited articles on the human genome and on boron carbide, and the *impact factor of* the *Japanese Journal of Applied Physics* doubled in 1995 by an article on gallium nitride. Even in highly cited journals, the majority of articles are poorly cited. As mentioned above, 50% of all articles are never cited! Sometimes an article will reach more interested colleagues if it is published in a journal with a lower *impact factor*, if that journal is read almost everywhere in the relevant scientific community.

General Scientific Journals

The best known general science journals are *Nature* and *Science.* Only articles of extreme importance whose implications radiate far beyond a narrow field are accepted there. Since the rejection rate of these journals is 95%, one can assume that the papers published there represent the crème de la crème of scientific articles, which is also reflected in the very high *impact* factors of these journals. If you are still a novice in the scientific scene, it is not necessarily recommended to try to place your first papers there already, unless your professor jumps up and down when he sees your results and suddenly becomes interested in participating in night shifts to get even more results faster.

The desire to get an article into *Nature* or *Science* is so strong among many scientists that Roger Highfield, science editor of the London Daily Telegraph, wrote: "Most scientists would kill their grandmother to get into 'Nature' or

---

[2] ISI also publishes a variety of other statistics and derived information. For example, for each field of knowledge and for each journal, one can find out the half-life of the citation frequency of an article. of an article. There is also information on the publication and citation rates of scientists, institutes and countries. A complete description of the possibilities available via ISI would go too far here.

'Science'." These are scientific journals with extremely high profiles and prestige (Fig. 4.2).

In all main disciplines there are other standard journals with a higher *impact factor*. This is especially true for journals that exclusively publish review articles.

4.2.4 Books

Books represent the most detailed and comprehensive type of scientific publications. In a book, that knowledge is systematically shown which can be found in over hundreds of publications distributed in different journals. A comprehensive and coherent picture is developed in a larger context, and the fundamentals are presented in an understandable way. One should always be able to learn something new from it.

Whereas a journal is concerned with the presentation of new individual results, only the recognised, established standard knowledge and fully confirmed findings find their way into a book. Readers of journals are assumed to be familiar with the basics. In contrast, a book should provide an instructive introduction to create access to the subject even for non-specialists. A book is in every way more detailed than an article and should help the reader to gain profound insights. The role of books is different from that of journals. A

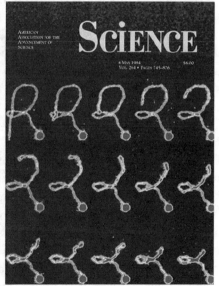

Fig. 4.2 Highly cited leading scientific journals

journal is about presenting smaller amounts of information in a field in motion, in a sense "fluid" knowledge. In contrast, a book reproduces "solidified" knowledge that is generally accepted. A book must provide comprehensible explanations and also present the background comprehensively. It is clearly more extensive and detailed than an article.

As with articles, different types of books can be distinguished:

Textbooks

Especially scientific publications in book form play a major role at all levels of teaching. In textbooks, the knowledge to be learned is presented in a systematic and didactically organised manner. The success of the learning process and the enjoyment of learning are not insignificantly influenced by gladly read textbooks. Scientific textbooks and monographs have an influence on the success of a student's learning and a scientist's acquisition of knowledge in self-study. In newly developing fields, after sufficient new generally accepted knowledge has been accumulated, it must also be systematically summarized in textbooks; consider, for example, anything related to nanoscience.

Monographs

These Books are written in a consistent and didactic style. Each author should be a proven expert in the relevant field.

The number of authors is small in order to keep the coherence of the presentation high, for which in case of several authors all chapters should be revised by each author so that no breaks in style become visible. Monographs should represent milestones in each field. They are written for scholars and advanced students who wish to enter a new field or acquire a deeper and more systematic knowledge in their field of work.

Edited Multi-Author Books

BooksThe books, coordinated by one or more editors, are scientific status reports. Each chapter has a leading specialist as the author, who presents the field in it as in a review article, but with an additional brief exposition of the basics. A thorough editor plans the content, coordinates the work of co-authorsHe takes care to avoid overlap and cross-referencing, and tries to bring coherence to the presentations so that the book gives a comprehensive and consistent overview. Sometimes an edited book is as coherent as a monograph. Often, however, such books are not consistent, for example, because of differences in terminology, and are therefore more suitable for scholars already experienced in the field than for beginners. Accordingly, such books are bought more by libraries than by individual scholars and students.

Reference Books and Manuals
In such books, data, techniques and the available knowledge in a field are compiled systematically and well structured. These books are considerably more extensive than the other types of books discussed, are much broader in content, in that entire larger areas of knowledge are covered, e.g. semiconductor physics or surgery, are thus also more expensive and are therefore essentially produced for libraries and laboratories. On the other hand, there are also smaller manuals on individual areas, e.g. Auger electron spectroscopy, which are also produced for the daily use of individual scientists and students.

4.3   Ethics of Scientific Publishing

What kind of material is suitable for a scientific publication? What is an ethically dubious publication? To answer these questions and to establish some general ethical guidelines for scientific publishing, the German Physical Society, the American Physical Society, IEEE, the American Chemical Society, the National Science and Technology Council (OSTP), the White House Office of Science and Technology Policy, and many other scientific organizations have developed and adopted a code of honor (see, e.g., *APS Ethics and Value Statements* 2002 [38]). This honor code attempts to define the culture of scholarly publishing and to ensure that honesty is a primary requirement of the entire business of scholarly publishing. The main statements of this honor code are as follows:

- Publish only new and significant results.
- Publish only original results, refrain from republishing of previously published results.
- The content of your articles is more important than their number.

Unfortunately, there is a tendency for quantity to take precedence over quality, because external research-funding agencies often measure the success of a project by the number of resulting publications.

American Physical Society Guidelines for Professional Conduct.

The Constitution of the American Physical Society states that the goal of the Society is the achievement of progress and the dissemination of physical knowledge. It is the purpose of the following resolution to advance the realization of this goal by establishing ethical guidelines for Society members.

The following defines minimal standards of ethical conduct that address some critical aspects of the physics profession:

* Research findings,
* Publication and authorship practices,
* Appraisal,
* Conflicts of Interest,
* Co-authorship,
* other findings.

Publish Only True and Confirmed Results. Do Not Invent and Falsify Data
While a prose writer is free to invent the content of his novels at will, scientific writing must be based on facts, i.e. truthfully report measurements or calculations. Also, results that might interfere with the picture being developed should not be suppressed. Therefore, accuracy is more important than speed of publishing. Theoretical results should have been thoroughly tested and experimental results should have been reproduced in at least a second measurement before they are published to eliminate errors. Outstanding results – if you are lucky enough to achieve them – require exceptionally thorough testing. An unexpected major discovery requires much more clear and conclusive evidence than results of routine investigations where the facts are already basically settled.

If you do accidentally publish an article that contains a fundamental error, then you should correct it with an Erratum and not give rise to further false speculation. speculation. If you repeat the error in subsequent publications to avoid losing face, you risk much greater damage to yourself in the long run when it is discovered.

Accept Intellectual Property and Copyright and Avoid Any Kind of Plagiarism
No researcherwho who writes a publication should use the results of other colleagues without making the origin clear. Otherwise it would be plagiarism. The grossest violation of the code of honour is the attempt to sell the results of other colleagues under one's own name. Such plagiarism will be outlawed.

Authorship Should Be Limited to Those Colleagues Who Have Made a Significant Contribution

Honorary Authorships or co-authorships of colleagues who did not contribute directly or indirectly to the generation or interpretation of results should be refrained from. Sometimes an attempt is made to include famous names in the list of authors so that this will get the article accepted by leading journals, or they are mentioned out of obligation or gratitude. On the other hand, however, it is not uncommon to list as a co-author the boss who managed to get the project funded and who contributed to the discussion of the results. as a co-author. But disciplinary power should never be abused to see one's name on many publications, as is unfortunately common in too many institutions.

In an extreme case known to us, the head of a research group, who worked exclusively as an administrator and no longer did any scientific work, forced his doctoral students and scientists to always carry him along as the last co-author of all their publications, thus trying to give the impression that he himself was strongly active scientifically. Once the number was large enough, he had the nerve to submit these papers collectively as a habilitation. Surprisingly, he managed to clear this hurdle, although there was isolated resistance on the doctoral committee. Sometimes outrageous or criminal behavior wins out. But do you want to take those risks? The example in Sect. 4.2.1 shows how extreme misconduct can ruin a scientist's career.

All Co-Authors Share Responsibility for the Content of the Article Over Which Their Name Appears

For the content of a publication is not the sole responsibility of the main author, but also of the more experienced colleagues who have provided scientific guidance to the younger colleague. Co-authors who have made minor contributions are primarily responsible for their part, but should endeavour to check the part of the other colleagues for correctness and accuracy to the best of their knowledge and belief. Any colleague who is not willing to take full responsibility for the article should refrain from co-authorship. This paragraph was added to the Honor Code… only in 2003, after the scandal described below became public.

4.3.1   Spectacular Cases of Misconduct in Scientific Publishing

Despite the noble goals of science, scientists are also human beings subject to temptation for possible personal gain. The history of science is replete with cases of both deliberate deception and careless publication practices that led to the appearance of sensational but incorrect results. The case of Jan Hendrik Schön is a spectacular case of a science faker who brought shame on himself, his colleagues, and his institute, Lucent Technology.

Jan Hendrik Schön

In February 2000, Schön, a young and promising scientist, published some surprising results. Schön and his partners at Lucent Technology (Bell Labs) had conducted research on molecules that are not actually electrically conductive and claimed that they could make them behave like semiconductors. This work was published in the leading scientific journal *Nature* and caused quite a stir.

At short intervals, almost weekly, further articles followed, of which a total of 14 appeared in *Science* and in *Nature* within only 1 year. Schön reported that he had succeeded in turning more non-conductors into semiconductors, and that he had been able to demonstrate their properties as lasers and non-absorbing devices. These results were revolutionary, and their implications for electronics and computer development seemed enormous. Flexible polymer electronics and nanoelectronics seemed to be breaking into new horizons. As one Princeton University professor opined, "Schön beat chemistry." Within 3 years, Schön wrote a total of 90 articles. In 2001, he received the "Breakthrough of the Year" award and numerous other science prizes; but most scientists considered this award ceremony only the beginning and expected the Nobel Prize soon.

But suddenly everything went wrong for the young prodigy. In April 2002, a small group of researchers contacted Schön's institute and expressed concern that all might not be well with Schön's data. Leading this effort was Princeton professor Lydia Sohn. She reported that one night, while critically reviewing Schön's publications, she and Paul McEuen of Cornell University found an unusual correspondence in the background noise in curves from different experiments. And some even showed the same data points. She said, "You would expect differences in the background, but the curves were identical." Skepticism was aroused.

McEuen became even more suspicious when he reviewed Schön's other published work. He found many duplicated curves from measurements of completely different systems. Apparently Schön used the same figures to tell different stories about them.

In May 2002, McEuen and Sohn informed the editors of *Science* and *Nature* about these inconsistencies. Afterwards they also informed Schön, his boss, the management of Lucent Technology and his co-authors that they would raise the alarm. Schön immediately explained that his measurements were correct and that he had only mixed up some curves for which he was now offering replacements. To *Nature, he* declared that he was "sure," and to *Science,* he said, "I didn't do anything wrong."

He could no longer show any measurement data because it could no longer be found and he had "reformatted his computer". He and his colleagues could not reproduce the suspect results in any way. The appointed investigating committee, headed by Stanford professor Malcolm R. Beasley concluded that he had freely fabricated his spectacular results on molecular electronics. He was found guilty of 16 out of 24 counts of scientific misconduct.

In his response to the investigating committee, Schön acknowledged errors but said it was not his intention to mislead anyone. He also wrote: "I have observed the various physical effects described in my publications, such as the quantum Hall effect, superconductivity in various materials, laser effects, and gate modulation in self-assembled monolayers, and I am convinced that they really occur, even though I could not prove them to the investigating committee."

He later said, toning down, "I wasn't faking, just anticipating future results."

Schön's answer failed to convince the authorities. Thus, the case of Jan Hendrik Schön became the most spectacular scientific scandal… in physics. It had also brought shame on the prize committees because outstanding prizes had been awarded for falsified results due to faulty decisions. The prizes were withdrawn, and even the prospective position of director of a Max Planck Institute vanished into thin air. Finally he was fired by Lucent Technology. Even the awarding of his doctorate by the University of Konstanz was cancelled, not because the dissertation had already been falsified, but because of the damage he had done to science. He was never given another chance to be employed as a scientist anywhere. Who would want to believe his results in the future?

During this time, about 100 groups worldwide were working to continue Schön's work. Suddenly, doctoral theses that had been started were hanging in the air, postgraduate young scientists had to worry about further funding for their projects, and the careers of some *assistant professors* and junior professors seemed at risk because their work included experiments based on Schön's publications. In this way, Schön not only ruined his own career, but dragged other colleagues down with him.

4.3.2 Other Cases of Scientific Fraud

Finally, for your amusement, I would like to list some other historical cases of high-profile, spectacular scientific fraud… of scientific fraud.

1912: The Man from Piltdown
In Piltdown, southeast England, archaeologist Charles Dawson excavated two ancient skulls and declared that this find proved the origin of mankind in Great Britain. It was not until 1953 that it was revealed that the "first Englishman" was from the Middle Ages and that a monkey mandible had been added to the skull.

1925: Adulterated Frogs
British biologist Paul Kammerer has been called the new Darwin. When he was accused by *Nature* magazine of manipulating images of frog evolution, he saw his career destroyed and put a bullet in his head.

1989: Cold Fusion
On 23.03.1989 Stanley Pons and Martin Fleischmann announced their discovery of "cold fusion". This was apparently the most spectacular discovery of the 1980s. But the enthusiasm fizzled out when it turned out that the interpretation insisted on incompetence and falsification. After that, Pons and Fleischmann's scientific reputations were ruined. They fled their laboratories and hid from public view. "Cold fusion" became synonymous with "scientific falsification." However, there are still some groups who dream of proving this phenomenal effect.

1997: German Cancer Research
An outcry of horror went through Germany when it turned out in 1997 that two prominent cancer researchers, Marion Brach and Friedhelm Herrmann, had literally sucked their fascinating results on cancer clarification out of their fingers for years. It could be proven that the results, at least those of the last

12 publications, were fictitious. This case became the biggest European scientific scandal to date.

2000: Caught Red-Handed

The respectable Japanese archaeologist Shinichi Fujimori was accidentally filmed reburying paleontological relics he had discovered elsewhere at his current dig site. He wanted to use the discovery at the new site as evidence of the broader historical significance of the find, a significance that the relics would not have had at the original site. After being caught in the act, he was released. All his previous excavations were now viewed with suspicion, and Japanese history books had to be revised.

2002: Elements 116 and 118

It was embarrassingIt was embarrassing when the famous Lawrence Berkeley Laboratory in California had to announce in 2002 that a scientist had freely invented the synthesis of the new chemical elements 116 and 118. The Bulgarian scientist Victor Ninov, who was responsible for the data analysis, had first tried to synthesize the element 116. He was lucky that shortly after the publication of the article, this element was actually detected. Because this had worked so well, he later tried again with the element 118. But here it was not possible to reproduce the detection anywhere. Doubts arose about the result. When the primary data were checked, the hoax was exposed and Victor Ninov was dismissed. However, the question remains: how could a single scientist falsify something as significant as the detection of new elements, when a large group of colleagues was involved in the work?

Hwang Woo-Suk

In 2006, the dream world of allegedly achieved cloning results built up by the Korean shooting star of cloning research Hwang Woo-suk suddenly collapsed when it turned out that all the spectacular results he had published were, unfortunately, nothing more than free invention.

4.3.3 What Leads to Misconduct?

While all the cases spread out so far are based on willful falsification, there is also still a wide grey area that can be characterized as careless work based on wishful thinking. Sometimes researchers become somewhat reality-blind when they measure something that does not fit properly into the wishful image. Also, competitive pressures occasionally give reason to publish results that are not fully confirmed prematurely, lest someone beat you to the punch and take credit for the idea or proof. If you must publish quickly, at least highlight the limitations and unresolved issues. That way, no one will accuse you of wrongdoing, even if it turns out later that the interpretation was wrong. A scientific reputation built up with much effort over long years of strenuous work can be destroyed overnight by deliberate falsification of results. Half-truths and lack of thoroughness should therefore be avoided at all costs.

4.3.4   Moral Guardians

Who is responsible for compliance with ethical standards in scientific publishing? First and foremost are the requirements for scientific honesty of the authors. The main responsibility lies with the lead author, who is usually the first author of a scientific publication and has done the bulk of the research. Older and more experienced co-authors, usually the supervisors or group leaders, must thoroughly review the draft publication and delete anything that appears to be incorrect, unconfirmed, or misinterpreted.

It is a useful practice at many institutions that each article must first be approved by the chief or institute director before submission to a journal, so as not to bring the institution into disrepute by publishing poorly. This also filters out useless, superficial or poorly written articles. In this way, the scientific prestige of the institute is maintained (Fig. 4.3).

It's not always easy for a reviewer to spot plagiarism… or falsification. Normally, a reviewer cannot easily answer the question: "Are these results credible, or could they be fictitious?" There are hardly any possibilities for a reviewer to check this in detail. Of the many reviewers who refereed the Schön work for *Nature* and *Science,* none were in any doubt that the data might well be made up rather than measured. The results seemed unique, but entirely possible. Even plagiarism is difficult to identify. If the reviewers however competent they may be, they cannot know all the articles and readily detect whether an author has simply copied from another previously published article or, in extreme cases, even sold it again under his own name. In order to exclude plagiarism, publishers use the iThenticate-program is used by publishers. This compares the text of a paper with all texts published in other journals or books. In this way, duplications are revealed and plagiarism or self-plagiarism is avoided.

Fig. 4.3 Beafeater © Andy Hooper/SOLO Syndication/picture alliance

The activities of the reviewers contributes significantly to the adherence to scientific standards. Therefore, reviewer activity must not be taken lightly. Reviewers who work carelessly and thereby abuse their privileged position in some way encourage the occurrence of misconduct in scientific publishing. If you are asked for a review and do not feel competent to do so, you should return the article to the journal or ask a more competent colleague to do so.

How should you behave if you receive an article for review from a colleague that you consider to be a direct competitor? There are two ethical and one unethical options: you can evaluate the article neutrally, without being influenced by competitive aspects, or return it immediately to the journal with the remark that there is a conflict of interest and that another reviewer should be chosen. and that another reviewer should be chosen. The third option is serious misconduct, on which a number of cases are known, specifically on articles in *Physical Review Letters*. Here, in the past, some reviewers viewed their opportunity to influence as a chance to delay the appearance of a competing article. Time-consuming additions or revisions were simply requested. In the meantime, the own article was written, also with the help of the guidance given by the other article. This made it possible for one's own group to publish the first priority-creating and most highly cited paper in the relevant field. Although this did not really contribute to the priority in terms of submission date, as this is crucial, it made them stand out as the first to publish.

To avoid anyone beating him to the punch, Chou, a Chinese scientist living in the US, had written "yterbium" instead of "yttrium" in the version of his article on the new superconductor $YBa_2 Cu_3 O_7$ submitted for peer review. When he received the galley proofs, he corrected one element. Surprisingly, another article appeared almost simultaneously with the wrong and non-superconducting composition. It was suspected that this article came from the reviewer.

4.4   Concluding Remarks

Although the compilation shown might give the impression that science is more often characterized by too weak resistance to seductions, hoaxes, and serious offenses, this is not typical. Scientific work is overwhelmingly characterized by honesty. Most scientists conduct their research in order to honestly discover new phenomena and gain new insights into the natural world, thereby contributing to world knowledge, rather than to gain dubious fame. Although scientists are not completely immune to self-interest, excessive ambition, and the greed for recognition, their vast majority behave correctly and are not driven by base motives. Mutual trust and respect are essential prerequisites for scientific results to be accepted and to contribute to progress.

5

Dissertation Writing

5.1   What Do I Want to Do and Where?

Let's assume you want to take the path to a doctorate, then the following question arises: At which institute, in which field and with which professor do I want to do my doctorate? You should not necessarily simply continue on the path you have with your diploma thesis. Of course, if you continue with your diploma or master's topic, you can start the doctoral phase with flying colors and continue to move in familiar surroundings, which is quite something. At some universities it is also possible to transfer directly to the doctoral pro-gramme after the bachelor's degree without a master's degree, which saves at least 1 year. However, if the dissertation is not completed, one is left without a Diploma or master's degree.

The more crucial question is which special qualification is likely to be in particular demand when you complete your dissertation and where you see the best future opportunities for yourself see for yourself. It is not just a matter of getting to the doctorate as manageably and quickly as possible. In this con-text, the possibility of doing a doctorate at a Fraunhofer-Max Planck or indus-trial research institute can be considered. There, research opportunities are usually just as good as at university institutes. The advantage of an industrial doctorate is the future security, i.e. the very probable transfer to a permanent position to continue the work under application aspects with a higher income.

For the selection of the supervisor it is also useful to have had previous contact with him. The role of the personal element should not be underesti-mated. If you are now toying with the idea of switching to another professor

© The Author(s), under exclusive license to Springer-Verlag GmbH, DE, part of Springer Nature 2023
C. Ascheron, *Scientific publishing and presentation*,
https://doi.org/10.1007/978-3-662-66404-9_5

or to another university straight away, then check the following aspects thoroughly beforehand:

- Should I build on the specialist knowledge acquired in the diploma phase or would it be better to change fields?
- Do I want a predominantly practical or experimental work or a more theoretically oriented work?
- Do I want to work more on the basis of my own research and analysis or more on the basis of literature?
- Do I want to concentrate on one method in a team of similarly oriented doctoral students? focus on one method or take the risk of developing something entirely on my own, which can be associated with time delays?
- Which topic do I want to devote myself to? Do I want to solve another minor problem in the safe environment of established research or plunge into the waves of an emerging field of research that is not yet completely certain of results? The former is guaranteed to lead to success, while the latter is more uncertain, but you can become famous by answering completely new questions for the first time.
- How promising is the possible doctoral topic? Is the doctoral field developing and does it promise to continue after the doctorate?
- Does the possible dissertation topic promise to answer the question within 2–3 years, or are there too many uncertainties, e.g. a very high time commitment for the construction of apparatus? It is not advisable to engage in tinkering with an uncertain outcome and a drastically delayed start to the PhD-worthy achievement, as numerous examples of failed PhD attempts show. After all, the doctorate is awarded for a scientific and not a purely technical achievement.
- Does the professor who would supervise the doctorate a good scientific, collegial and human reputation? Usually, the most prominent professor has the largest and best-equipped research projects.

It may well be significant whether the supervisor is of such importance in the scientific hierarchy that he or she can also become a reviewer, preferably the first reviewer, and thus positively influence success in passing the defence hurdle. He can then, under certain circumstances, replace the at least two other reviewersone of whom should be external and who are not so close to the work. And besides, one then has at least one committed advocate in the defense.

These questions you can partly clarify from a distance. But there are also some other questions that are better elicited on site. Some things you can better assess if you talk to the professor and the other doctoral students on site:

- How good would the care be, and how personal can the relationship be?
- Does the supervisor support his doctoral candidates in case of difficulties or does he leave them alone at the mercy of the problems?
- How do I get along with the counselor? Are we on the same wavelength?
- Will the supervisor support me even if I get better results than expected or then hinder me as a competitor? This can happen with supervisors who are aspiring younger scientists, while it is less likely with detached older professors.
- Does the professor support his doctoral students not only in achieving good research results, but also in finding good jobs afterwards? For example, in Würzburg there was a very well-established older professor of medicine who had placed all his post-doctoral students as professors or clinic directors.
- Can I establish a good relationship with the other doctoral students already present and expect support from them? The emotional component of feeling comfortable at the place of work should not be underestimated.
- Does the environment suit me (location, buildings, rooms, working conditions)?
- How long did it take the group in question to complete their doctorate in the past, and what was the success or drop-out rate??
- How far does academic freedom go and what is the compulsory attendance?
- Do doctoral students in this group have the opportunity to present their results internationally and have outstanding achievements been made?
- Is the professor/supervisor part of a scientific network in which there is lively scientific exchange with other, also international, groups? For example, are visiting scientists frequently invited and does the exchange also go in the opposite direction, so that you also have the opportunity to get to know other institutions and later to be able to make more confident decisions about your next place of work? As a doctoral student, do you have the opportunity to build up your own international network? in which a position may later open up for you? Through such a network, many tasks can be solved in a more complex way by incorporating the methods cultivated by other institutions. The interdisciplinarity of research is increasing worldwide. Therefore, interdisciplinary transfer of methods is very useful.

5.2   Selection of the Topic

Once you have decided on a research direction, the location and the doctoral supervisor, the thinking and discussion about the concrete dissertation topic begins. Possibly, your new supervisor will present you with the open questions in the chosen rough research area that could become dissertation topics, or make a concrete proposal right at the beginning.

Then, take the time to think about it more deeply and learn about the state of knowledge in the field before making the decision. To get started, you may want to read relevant review articles basic publications and the most recent articles in this field can help you to get a more concrete overview. Perhaps your future supervisor will already give you some guidance on this. However, you can also find them yourself on the Internet and in the online versions of the journals by searching for keywords.

It may also be that the current state of knowledge in the field is summarised in a recent book, which is a useful introduction. Occasionally, open questions are presented at the ends of chapters. This could also provide food for thought about possible dissertation topics give. However, care should be taken here and enquiry with the author concerned is useful to ensure that these topics are not already being worked on in his group. This is because the same results twice are only recognised as a dissertation in the first solved case.

Also talk to friends and colleagues about a possible dissertation topic for yourself. However, do not focus exclusively on the topics suggested by your supervisor. You may have come up with another interesting idea in these discussions or while reading the articles.

Try to be clear about what major new questions you could address in a dissertation of your own. If you find appealing questions that you are willing to invest the bulk of your time in answering for the next 3–4 years, the next step is to consider how the broad topic could be broken down into individual, smaller questions to be addressed could be broken down.

It is not essential to accept the first proposal of the supervisor without any modifications. In my dissertation, I also did not accept the first proposal, which would only have meant a continuation of a topic already established in previous dissertations with the existing experimental methods, without the scientific novelty being essential. For the professor, a certain extension of his data base by a new measuring servant would have been useful, but I did not see in it the solution of a new topic worthy of a doctorate. After a thorough study of the literature, I decided on a different topic. This one I had to push through in a long personal discussion against the scepticism of the supervising professor. But finally it led to the Gustav Hertz Prize of the German Physical Society for the best physics dissertation of the year 1980.

5.3   How Do I Organise My Work?

You have now clarified all these issues and a number of others that are important to you in advance and have decided on a place to do your doctoratea supervisora group and the topic decided. Now the actual work begins.

The requirements for the different qualification works can be compared to a race over different distances:

- medium distance (3 km): master thesis,
- long distance (10 km): dissertation,
- Marathon (42 km): Habilitation.

Accordingly, the average periods are 1, 3 and 6 years.

If you have now decided to do a doctorate, then the further responsibility for achieving this goal lies essentially with yourself, even if your supervisor and colleagues still try to influence you to work responsibly. You will have to show great perseverance.

It is the requirement for an achievement worthy of a doctorate to have solved an essential scientific problem anew for the first time and independently, as has already been stated above. This goes far beyond the requirements for a diploma thesis or master's thesis, where it is a matter of solving minor scientific problems under guidance.

However, a doctorate is not the same as a habilitation, which must involve the new development of a major scientific sub-field including successful teaching. According to the smaller scientific scope of a dissertation compared to a habilitation it cannot be the goal to solve all the world's puzzles in this work. At some point, the doctoral candidate must develop the courage to fill in the gaps. Good students in particular reach this insight very late. You must aim early for a well-rounded whole, rather than a complete treatment of the subject that slays all possible questions. Keep in mind that in the end, most universities will not allow you more than 100 pages to present your findings in the dissertation paper.

5.4   Time Planning of a Dissertation

Do not rush into the planning of your dissertation. Consider beforehand how best to organize your work in order to submit it successfully after 3 years. Draw up a detailed plan in which you break down the large topic into many smaller manageable individual topics. Discuss this concrete dissertation plan with the supervisor and other colleagues involved. Modify the plan according to the competent advice of others. Finally, anchor the whole thing in terms of time by specifying which goals are to be achieved and when. Also coordinate the timing with the supervisor. In this way, you have a basis for periodically being able to account for smaller successes in front of yourself.

Now you need to concretely plan the experiments or other necessary work. Start soon with the necessary preparatory work. Then you can soon get started with your new demanding work.

You may be lucky and make progress as planned at first. But then things suddenly get stuck. You don't achieve the expected results straight away, the repair of the apparatus takes longer than expected, the answers to the distributed questionnaires are not received, the subject of the investigation proves to be more recalcitrant than initially thought. They can't think of a way out. It often happens that a doctoral student loses motivation... often loses motivation because there is no sense of achievement at all, as he or she works like a mole to complete a very complex task and the visible results only appear after a long period of time.

If things do not progress as desired, sometimes almost depressive feelings of being lost or lacking competence arise. In such a case it is good to have friends among the other students and colleagues who help you or lift you up again. In case of seemingly unsolvable problems, it is helpful if you can count on the support of the other graduate students or other colleagues and the supervisor or professor. If you can't seem to get out of an impasse at all on your own, talking to competent outsiders often saves the day.

To avoid this negative experience, it is helpful to break down the task into smaller, manageable and concrete sub-tasks as discussed above. When working through the sub-tasks, you will periodically see successes and know where you stand. The higher goal is always reached only through many smaller intermediate steps. For this purpose a time schedule is a valuable help. If you continuously work through your task list and produce the planned results, you will not drastically exceed the time allowed, as happens to far too many doctoral students who only get their work done in an extension phase. Take your plan in for a monthly fulfillment check. This way you will see where you still have to work particularly hard.

Because progress in scientific work can only be seen after a long time, Albert Einstein loved to chop wood. Because then he saw the result of his work immediately.

In order to avoid unnecessary loss of time in case of unexpected problems, we recommend you to develop a complex, parallel and not only linear way of working. Approach different issues in parallel, then you will not have to wait long if things do not progress as desired in one area. Be prepared to go through phases of result evaluation or literature study in waiting periods. Perhaps you can also write the first drafts of parts of your dissertation in waiting periods for completed parts of your work or set down completed sub-areas in initial publications.

Do not practice *learning by doing* aloneas most doctoral students do. Try to acquire systematic knowledge about your research topic and also read a text-book, a scientific monograph or work through an edited book thoroughly as a scientific status report. After all, the more profoundly you are familiar with the subject matter, the better you will be able to classify the newly achieved results and develop ideas for further action.

It is usually a very lengthy process to generate the results needed for the dissertation. Either extensive experimental work is required, the evaluation of studies and questionnaires or time-consuming theoretical work and/or calculations. It also involves extensive study of the existing literature. If you have the firm goal of producing the necessary results in the given time frame, this means for usually 3 years of very intensive work, setting aside many personal and leisure interests. Approach this goal with a firm determination to succeed.

Make intensive use of the opportunities offered to carry out your own experiments, or carry out your work at your desk and computer with a high level of concentration. Experimental work on existing apparatus usually requires coordination with other interested parties and technical assistants. But colleagues can also help. A good personal climate is conducive to results.

Also try to work in a creative environment to be able to continuously develop good new ideas. If your study room at the institute is too crowded and noisy, you can also arrange to work undisturbed in the library or at home.

Don't forget to have enough sleep and relaxation during the intense work periods to avoid burnout.

Perhaps you will also experience that the particularly intensive occupation with a task, in which you become absorbed and with which you identify completely, leads you to develop a feeling for the scientific problem. This then helps you to make seemingly complicated scientific decisions quickly and intuitively. This can greatly stimulate the progress of one's own work.

But also try to keep the general overview and not just bury yourself in detailed tasks like a mole. Make an effort to be self-critical of your results and approach, as if from an outside position. This can lead to a more effective work style.

5.5   How Do I Document My Results?

Make it a rule to keep a record of all the results you have achieved: in a constantly updated file on the computer (better still also backed up on a memory stick or an external hard drive), handwritten in a lab book or notebook in which you record your progress in detail. It should not just be a loose-leaf collection that might later get jumbled or lost. Any contradictions that may arise are easier to clarify this way. Perhaps at an advanced stage, results that were initially considered insignificant may also appear useful. Keeping an organized record of all results will facilitate their later use. Keep in mind that the recorded results can also be used as legal evidence when priorities are at stake. For example, the long-running dispute over the granting of the patent for the invention of the laser to Gould was decided in his favor only on the basis of the certified laboratory protocols, which abruptly made him a multimillionaire, since all laser manufacturers had to pay retroactive royalties.

Now that you have been working on your topic for some time and have carried out many investigations or calculations, it is time to document what you have achieved in writing in publications. By writing a publication, you condense and generalise your results, compare them with published results of other scientists and classify your work. You thus reach a new level of knowledge.

Most scientists have realized that writing dissertations and publications is not just a matter of writing down the results, but a highly creative process .... Because when you try to present everything coherently and convincingly, you often become aware of cross-connections become conscious. One also becomes aware of which partial results may still be missing and still need to be generated. In this case, further investigations may be necessary in order to be able to present the problems as comprehensively solved.

A certain number of publications is expected as recognition of the achievements by other scientists before the submission of the dissertation expected. For this purpose, the information provided in Chap. 3 and the help of more experienced colleagues will be useful to you. If you now start to publish your results, you will not only achieve a new quality of your work, but at the same time you will start to take a place in the scientific community. Moreover, you can make good use of all that you have already published in articles, block by block, for your dissertation. You will certainly also be able to add a few research reports will be added.

Depending on the career goal, publications have different values: For a research career, it is important to have many good publications in renowned journals. But if the goal is a position in industry, then the publications are only important for achieving the doctoral goal, and this should be done as quickly as possible. In this case, patent applications can be of greater benefit.

5.6   Writing the Dissertation

5.6.1   Basic Requirements for a Dissertation

If you have now actually achieved all the planned results to your own satisfaction and that of your professor, so that you have the good feeling that it is time to summarize everything in the dissertation, then you can first of all breathe a sigh of relief, especially if it has worked out in the allotted time. If you have already compiled your main results in publications and have to rearrange them more or less block by block in order to present an overall picture in the dissertation, then writing a doctoral thesis of about 100 pages is not too difficult a process. The discussion has already been worked up, and all the figures are ready. Also, hardly any doctoral reviewer will doubt what journal reviewers had already accepted in advance. Now, put the pieces together harmoniously to present your new findings in a condensed and convincing way. Justify all statements as far as necessary.

In most cases, however, the attempt to summarize the work reveals which gaps have remained open; and further work is often still necessary when one thought the end was already within reach. Sometimes one can get the feeling that one is only approaching the longed-for goal with infinitesimal steps. Now once more stamina is required. Also, you should have already collected in passing all the literature required for the interpretation to be discussed. Therefore, it is important that you continuously take the time to follow the scientific literature in your field, no matter how pressing the day's tasks may be. It is recommended that you set aside one afternoon each week for literature study.

What should a dissertation thesis contain? It must be an independent scientific achievement that independently solves a new scientific problem.

Novelty and scientific content of the work are the decisive criteria. The prerequisite of the work is that the author has investigated which questions and problems exist in the subject area he or she is dealing with and which approaches to solving them are already known in the literature. The development of own original solutions is crucial. If it should turn out that, in ignorance of the doctoral candidate or his/her supervisor, the apparently new solution has already been published by other scientists, then this merely reproducing achievement is no longer worthy of a doctorate. Under no circumstances should a scientific or engineering dissertation be merely a summary of secondary literature (commentaries, definitions, monographs, articles, etc.).

In general, the following applies to a dissertation: the author should not only report, but investigate, explore, find new things and explain them.

5.6.2 Formal Structure of a Dissertation Thesis

A dissertation should be proportioned in the same way as an article. The focus has to be on the presentation of results and interpretation. Everything said in Chap. 3 should also be applicable to the writing of the dissertation. In addition, note the following points:

- Pay attention to a logical didactic structure and a clear structure of the dissertation. They are the be-all and end-all of any thesis. From the structure or outline of your work, your own concept and the logic of your thinking become clear. Your concept of approaching the topic and not only the content of the work is your *signature,* which gives your work its independent character.
- Choose a well-structured outline with no more than three levels of numbering. Think of the individual headings as questions. The following chapter must provide the answer.
- Make the presentation of the results coherent and not fragmented. Formulate the discussion in an understandable way. It is best to discuss the results immediately after they have been presented.
- Often the mistake is made to write down a lot of things more or less mechanically. Instead, the work should take its own approach and approach the topic logically and systematically.
- In some dissertations the discussion is somewhat meagre, and it is only stated that something is *so,* but not why. The crucial thing, however, is not

just the result, but also the reasoning behind how one arrived at this interpretation.

- Provide an introduction and clearly state the task and topic at the beginning. The motivation for your work must become clear here by elaborating the still open, essential questions in the existing body of knowledge. Present the basics in a short form.
- In humanities dissertations and some medical dissertations, a hypothesis is prefaced, which is proved by the following exposition. However, this is not common in the natural sciences and engineering.
- This should be followed by a methodological or experimental section in which you explain how you obtained your results. If you have developed a new method yourself as part of your work, explain it in more detail than the known standard methods used. These should be treated in an abstract-like manner.
- The main part of the paper, both in content and scope, is the presentation of the results and their interpretive discussion. Present and discuss your results with the necessary breadth, drawing on what has already been published by others in your field. Be aware that in terms of content and scope, the focus must be on your own results.
- Draw conclusions at the end of the chapters and the thesis. These conclusions must go beyond the narrower discussion of results and make clear the wider significance of the work presented.
- Make sure that the focus is appropriate. In particular, you should summarize passages that are not directly related to the paper and limit yourself to what is relevant to the issue under investigation. By dealing with marginal questions and important areas in detail, you show that you have an eye for the essentials. For important key points, however, you must go into depth.
- Literature references are important to compare and validate your own interpretation with that of other scholars. Make sure you have included the essential relevant literature in the discussion. As a rule, a dissertation should contain at least 100 citations containbut usually considerably more. Often a glance at the bibliography often provides information about how thoroughly the author has worked and how familiar he is with his subject area. All results and thoughts that do not originate from the author must be made clear as such and their source must be indicated. Do not focus exclusively on current literature, as unfortunately most young scholars do out of convenience; consider the older basic literature as well. To be on the safe side, you can do a full-text search in the electronically available archive section of all international scientific journals using key words before you finish

writing. Citations from relevant books also show your good literature overview and basic understanding.

- Avoid anything that could give the impression of plagiarism. Make it clear what you are taking from others. At first glance, it may look good to present essential thoughts and results of other colleagues as your own. But the result of an in-depth examination could be devastating for the plagiarizing young scientist and lead to non-acceptance of the dissertation.
- Format Cite in the required style, most often the international journal style is required. Follow the style of your field or the regulations of your faculty. It is also helpful to have looked at some dissertations of your own department beforehand.
- Make sure you stick to the specified number of pages. For dissertations in the natural sciences and engineering, this is usually 100 pages, while 200 to 500 pages are usually accepted for papers in the humanities. If you have only a few results to discuss, avoid filling the excess pages with material that has no logical relation to the actual topic of the thesis. This will only irritate a reviewer.
- It does not necessarily have to reach 100 pages. There have also been much shorter dissertations. Edwin Hall's dissertation was only four pages long, describing the Hall effect he discovered and later named after him. This result was so spectacular that it was accepted as a brilliant dissertation.
- A summary towards the end of the dissertation in connection with an outlook on necessary further work is very useful. Possibly, your dissertation can lead to another one, or you can continue to research the problem in depth afterwards in the direction of a habilitation.
- Precede with a list of abbreviations and symbols if you often have to use abbreviations or symbols for the sake of brevity. Explain all abbreviations and symbols in the continuous text and write them out additionally at the first mention.
- At the end, do not forget the acknowledgement, in which you thank for all the support shown and the boss at least for the task.
- The affidavit about the independent preparation of the work must not be missing at the end.

5.7   Defending the Dissertation

Prepare your defense thoroughly. To avoid possible fear of not knowing everything, be aware that no one knows as much about your topic as you do.

You should present your results in excellent slides. Detailed instructions on how to do this were given in Chap. 2. Your defence presentation should not only be impressive in terms of content, but also visually.

Also keep in mind that your speaking and expressive skills… must also be excellent. Preliminary exercises, e.g. at a group seminar or in front of friends and a video recording will help. Practice the well-prepared speech several times beforehand, and speak freely. You can enhance your work by giving a well-delivered talk and devalue what you have worked hard to build up by giving a poor talk. To be able to react with presence of mind, approach this important event well-rested and not under sedation.

Make sure that you keep to the given time frame, usually 20 min. For the presentation you should take the presentation hints given in Chap. 2 to heart, as you have already done for your conference presentations. When preparing the content of your defence speech, be aware that, especially in the case of a dissertation defence, the audience is often very heterogeneous, ranging from employees who know every detail to professors from a completely different field. There should be something for everyone in the lecture, without the seriousness of the presentation or the dignity of the event suffering from too much loosening up of the lecture.

Do not forget to appear appropriately dressed. The dignity of the event, as the most important scientific event of your career so far, should also be reflected by your clothing should be reflected in your attire.

Avoid anything that might cause irritation to the audience and especially to the evaluators, such as hand in pocketfrequent scratching or constantly circling the pointed segment of the slide with the laser pointer.

Keep a close eye on the expert witnesses during the defence presentation and be quick-witted if they frown or shake their heads.

Surely you can answer perfectly all questions concerning your direct work. However, if there are questions about peripheral areas with which you are less familiar, try to give sensible answers to them as well, perhaps skilfully diverting to other related questions about which you are able to say something competently. You may well note that this topic was not the subject of your work.

If, contrary to expectations, there are aggressive questions that snub you, remain friendly and competent even in such a situation. You can answer with a counter-question, which will then usually break the point. The answer could also be redirected to a related area.

It is not a good idea to make arrangements with friends to ask questions that you are then sure you can answer well. If this becomes known afterwards, it is more harmful for you than to react with uncertainty to unexpected questions.

If you have now also survived the discussion after your excellent defence speech, you can breathe a sigh of relief and invite everyone involved, or at least your closer colleagues and friends, to your doctoral celebration. They will then congratulate you on the successful completion of your lengthy work.

5.8   Purchased Doctoral Degree

Not to be taken seriously are guide books that describe how to get a dissertation in the shortest time and with the least effort.

At the end of this chapter I would like to mention that apart from the serious, time-consuming and labour-intensive way shown, there is also a dubious, very short way to get a doctorate. A number of scientifically not to be taken seriously institutions, which have acquired names of obscure Latin American universities, offer purchasable doctoral degrees for 3000 to 5000 EUR, without having to write and defend a dissertation. Such an almost genuine-looking doctoral certificate can, under certain circumstances, lead to the entry of the doctoral title in the identity card if the proper examination is omitted. This title, however, does not stand up to scrutiny by a scientific board. An employment or career in research cannot be brought about with it.

6

Career Planning

After all, you already made a far-reaching career decision when you chose to study. But when your studies are about to be completed, a doctorate is almost defended or a postdoc position is coming to an end, you find yourself at a crossroads-position is coming to an end, you are at a crossroads and need to make further fundamental career decisions.

6.1 After Graduation

You're now thinking about whether you want to do another PhD... or whether you should look for a better-paid position in industry. Why not try to bring a certain continuity into your further professional development, so that one stage can build on the other without you having to manoeuvre your way out of dead ends later on? After you have completed your diploma or master's degree, you need to answer the question: Does a doctorate bring me anything at all? If so, what do I want to achieve with it, a research career, perhaps all the way to professor, or a better position in business, possibly as a highly paid management consultant?? However, a management consultant should only be someone who has already successfully managed a company. Do I even want a career without a scientific career from graduation to retirement?

If you want to continue doing research as you did in the diploma phase or if you are planning a university career then the answer is clear: do a doctorate. In many disciplines, a doctorate is a must: 90% of chemists do a doctorate and can only find good jobs in industry with a doctorate to find good jobs in

C. Ascheron, *Scientific publishing and presentation*, https://doi.org/10.1007/978-3-662-66404-9_6

industry. For physicians, the doctorate at many universities is so simplified and flattened in the scientific claim that it can be started in lower years of study and the study is completed as a Dr. med., although the title is awarded only after the state examination. Lawyers and psychologists appear better in public if there is a Dr. in front of their name. On the other hand, the Dr.-Ing. is sought much less frequently, because engineers get good jobs in industry even without a doctorate.

The doctoral phase is in most cases another financial lean period. In the natural sciences and the humanities, a doctoral student receives only 50% of the TVöD salary (TVöD is the collective agreement for the public service). The situation is better at universities and universities of applied sciences with doctoral degrees in the engineering sciences. There the doctoral positions are full positions.

Can you or do you want to afford another low-income phase? You have already lost a lot during your studies: If we assume a monthly cost of living of EUR 1000 and a loss of earnings of about EUR 2000 compared to people starting their careers after an apprenticeship for a period of six years (average period of study at German universities), then your studies cost you about EUR 220,000 in lost income. This does not include costs for books, computers, other teaching material, tuition fees if you exceed the standard period of study or fees for any private universities are not even included. So let's add another 10,000 EUR or so. At an American elite university, where the tuition fee is already 25,000 EUR per year, it gets even more expensive. Even though your parents might have covered the basic costs, you're still left with a drastic loss of income compared to non-students who started working at the age of 18. If we assume that you will work for about 40 years, then the apportioned loss of income over the years will be about 5500 EUR/year. Now, if your monthly salary as an academic is about 450 EUR higher than that of a skilled worker, which is very likely, then the study investment has paid off. So don't consider the delayed start of your career due to your studies as a financial disadvantage. However, what is more difficult for university graduates to make up for is the financial advantage of graduates of universities of applied sciences, who much more often complete their studies in the shorter standard period of study.

If you stay at an institution of higher education, then you can read your income development from the TVöD tables. In most cases, things look better for scientists and engineers in well-paid industrial positions, who do 15% better on average in the statistics, as can be seen, for example, from the annually published income comparison for *physicists in* the *Physik Journal*. However, in industry the work pressure is higher and the academic freedom is low. There is also a spread of income that varies greatly with industry and company size. Supply and demand, and how well one sells oneself in salary negotiations, have a significant impact on more freely negotiable industry salaries.

So the following questions remain to be answered in order to decide whether to do a PhD or work in industry or industry:

- What is my career goal? Where am I drawn to? Am I more of a researcher type or a practitioner? In a university job, there is indeed independent research and hopefully satisfaction through excellent and internationally appreciated results. However, the latter need not be the case and can instead involve long dry spells with no fulfilling moments, with little practical benefit to humanity from one's work. In an industrial activity with smaller, manageable, shorter-term projects to be completed in manageable periods of time, where the immediate benefits become clear upon completion, the sense of fulfillment may be greater. The decision to pursue a career in university research involves teaching. One should decide early on whether one can give good talks and lectures, whether one wants to be involved in mentoring junior researchers, and thus whether one can become a good university teacher.

The result of the diploma thesis has only limited significance for this decision. After all, good diploma grades and the will to perform are possibly necessary, but probably not sufficient criteria for a doctorate that is not only formally successful. While the diploma thesis was done under supervision, the doctoral thesis is about independent research. Initiative, stamina, organisational talent and the willingness to continue learning are required for this.

- Will the doctorate something professionally? The desire to have a doctorate in front of one's name cannot be dismissed out of hand. There may be many more or less understandable reasons for it: from one's own aspirations to complete one's university career, to the desire to satisfy one's parents' pride and ambition, to the desire for a doctorate. It may have many more or less obvious reasons: from one's own aspirations to complete one's university career, to the desire to live up to the pride and ambition of one's parents, to corporate cultures, for example in consulting firms, where the title facilitates access to clients. Quite apart from this, the title of Dr. med. is almost mandatory for doctors to be taken seriously by patients. Here, it is often only the Dr. on the business card that matters and not the content or quality of the dissertation. If it is only the title itself that matters, then you should try to get through the qualification phase as quickly as possible, and you will hardly care whether the dissertation is defended *summa cum laude* or only *rite*. In contrast, for a planned university career, the doctorate is a must. Here, the quality and grade of the doctorate matters a lot.

6.2   Postdoc or Work in Industry?

Now you can enter professional life well prepared or take on the uncertainty of an academic career. At this point, some people also think about setting up their own business and taking the risk of founding a company with start-up capital raised somewhere, in order to implement the knowledge they have gained for their own benefit. Perhaps you feel you have done enough research at an academic institution and now want to work somewhere else entirely, possibly in a company. Or you may have taken such a liking to gaining new knowledge and the freedom of academic life that you want to continue working at a research institution in a postdoctoral position. -position at a research institution, either at a university, a Max Planck Institute or another research institute, including industrial research institutes also offer opportunities. Non-university institutions such as Max Planck Institutes generally offer better research opportunities for a postdoc.

This would also be the time to consider a postdoc position abroad. Gaining international experience can never hurt. A stay abroad should advance you personally and professionally. A positive side effect is a better command of a foreign language. Especially if you return from prestigious institutions in the USA with good results, you will have good chances for further applications in Germany. If you do not find a suitable advertised position – funding is always the main problem – you can also apply to the German Academic Exchange Service (Deutscher Akademischer Auslandsdienstthe Humboldt FoundationEU Research Promotion Foundation or the Fulbright Foundation (for USA), if there are possibilities for scholarship applications there are available. In Japan, attractive positions are offered for creative postdocs. The organizing institution there is JSPS (Japanese Society for the Promotion of Science), which also has an office in Bonn. In Europe, there are further opportunities for postdoc positions in EU programmes. You can also find attractive positions at leading universities in the UK, France, Italy and Spain, often as project positions – and the journey home is not quite so far. But you may be able to better complete your specialist skills even more effectively in a less famous, smaller institute. In the US, after four postdoc years, you may have a chance to move into a *tenured position*which is a predefined university career path leading to a professorship... to a professorship. At industrial, Max

Planck, Fraunhofer- or other research institutes, it is not uncommon to have the opportunity to transfer to a permanent position. However, you should be aware that a postdoctoral position is in most cases only a further is in most cases just another transitional phase, after which the uncertainty and job search continues. Many scientists just shimmy from one temporary position to another into their 40s. However, if someone has a family, other criteria such as financial security may become more decisive than the joy of a free life as a researcher.

The same criteria apply to the selection of the country and the institution or professor in whose group you intend to work as discussed above for the choice of doctoral institution. The main question is: Do I want to build on the knowledge acquired during the doctoral phase, or do I see a more attractive new field that promises better career opportunities?

Don't get involved in uncertain and very long-term projects with a temporary position. You need to cash in on successes to either get an extension or a good permanent position at the current facility or elsewhere. Having successfully completed several shorter projects helps more than having to leave a great long term project unfinished. Remember to produce many good publications in reputable journals during this period. This is what you will be judged on, and it can improve your application prospects …a huge boost. Establish yourself as a respected member of the research community in your field.

It is not wrong to use these four or so years of research to work on a habilitation. When the postdoc period comes to an end, you can apply for a professorship. For someone who has only worked in academia up to that point, a university professorship is then open in principle. However, if you have worked at an industrial institute for at least three years, you can also apply for a professorship at a university of applied sciences, as this does not require a habilitation.

In the postdoc phase, many people are confronted with the task of having to finance part of their work, or after a while possibly even their entire salary, through project funding. So they have to learn how to write and defend them. Try to benefit from other colleagues experienced in these matters.

But perhaps after your doctorate a job in industry might be closer to your heart. If you have already made this decision at an early stage, it can be very useful to make contact with the desired company before you start your doctorate and to try to align the dissertation with the research interests of the company in question while you are still at university, i.e. to choose an application-related topic or to work on the dissertation at the industrial institute straight away. Then a smooth transition is possible and the question of employment can already be clarified in advance.

Otherwise, there are many uncertainties: Are you too highly qualified so that you are entitled to a higher salary that they don't want to pay, or are your specialist qualifications more likely to be seen as a nuisance? In Japan, for example, companies are reluctant to hire PhDs because they have developed a more independent way of working than graduate students fresh out of university and are considered less malleable. But you may also be welcomed with open arms as a specialist in a sought-after method. In any case, it is advisable to make contact with companies in good time before completing your doctorate.

6.3   After the Postdoc

6.3.1   Job Selection

It is to be hoped that you will find a good permanent position either at the postdoctoral institution or elsewhere with the experience you have gained. But you can also plan to continue climbing the academic ladder, either in a postdoctoral position or a junior professorship. The latter would guarantee income and professional development for another six years or so – but with no guarantee for the time after that. This is because most junior professorships do not offer the option of conversion to a permanent professorship later on, even if the chances of applying to other universities are not bad afterwards. And it is uncertain whether you will find a professorship with your habilitation or whether you are overqualified for other applications. What is certain is that you need a qualification phase of at least 12 years, usually even more, from studies to doctorate to habilitation. It is difficult to guess whether habilitation or junior professorship are the more successful way to obtain a professorship later on. Although Germany has announced that it will abolish the habilitation as a prerequisite for appointment in the medium term, this is currently the safer path. Incomprehensibly, many universities do not allow junior professors to submit a habilitation. At many universities, an *associate professorship* at a foreign university or a management position at an industrial research institute.

The advantage of a junior professorship compared to a habilitation position is not only the use of a nice-sounding title, but also the greater freedom to conduct independent research earlier, to acquire project funding and to be allowed to attend all faculty meetings. However, a postdoctoral fellow can often concentrate more on research, as he or she can take advantage of the department's existing infrastructure and is often less burdened with administrative tasks.

But in both cases, the uncertainty of another application phase begins again. If, after the junior professorship or habilitation, it does not work out soon with a professorship, only a few habilitated scientists or expired junior professors can be kept afloat in Germany for a limited period of time by the Heisenberg fellowship in Germany for a limited period of time. A visiting professorship or guest researcher position abroad may be the only way out if there is nothing to be found at a research institute in Germany.

6.3.2 Interview

When applying for a postdoc position as well as for an industrial position or professorship, job interviews are common. This is where the preliminary decision is made as to whether you will be accepted as an applicant for the advertised position. So try to make a good and competent impression. Go there as well prepared as possible. Find out about the institution, the focus of the work there and try to gather more peripheral information. This is another way of expressing your strong interest in the job. The more informed you can appear, the better your chances will be.

Think about possible questions and the best answers to them. Typical interview questions are:

- How has your professional development been so far?

  Present your professional development in a comprehensible and targeted manner. Do not give the impression that this position is the last resort for you. Try to show that your career has created the prerequisites for the job you are seeking.

- What is your experience and knowledge?

  Don't put your excellent knowledge too prominently. Otherwise you may get the impression that a competitor is appearing. It can be supportive to emphasise that you are always ready to face new challenges and that you can quickly familiarise yourself with new areas of activity. Dealing with realities in the right way is just as important as being clever.

- What are your personal strengths or weaknesses?

    No one seriously expects a truthful self-portrait in response to this question. Here it is necessary to show that you have the qualities required in the expected work environment. However, do not stray too far from reality when describing your outstanding qualities. Provide arguments that speak for your attitude.

    The question about personal weaknesses is not about the pure truth, but about an eloquent and yet self-critical sounding response. As a young professional, you can mention, for example, the lack of practical experience or the desire to solve every task perfectly. Show that you have the ability for realistic self-assessment. You can definitely score points here with appropriate preparation.

- What have been your biggest successes or failures so far?

    Name a research task that you have done brilliantly, for example, in a project, and point to published results. As a negative experience is more suitable something in which you yourself were not to blame for the failure.

- What would you do if we hired you?

    Don't just stick to your own area of experience, which may not be in demand which may not necessarily be in demand. Show yourself to be open to new challenges. Prove that you have always been able to quickly take on new tasks. Emphasise the possibility of your immediate readiness for action based on your experience to date.

- What would your research plan be for the next few years if we were to appoint you to this professorship? Can you outline your research plan?

    Go into more detail on the subject, and also say something about collaborations and – especially in the experimental area – about the amount of funding required for the endowment of the professorship. Think about this possible answer carefully beforehand. If you want to take the research group too far away from its previous field of activity, resistance is likely to arise. If, on the other hand, you do not intend to introduce anything new at all, ask yourself what new quality your appointment would bring. If you have inadequately high expectations when it comes to finances, you will worsen your chances. Conversely, if you are too modest, you will limit your own opportunities. So it is better to inform yourself about the possible framework beforehand.

- How is your ability to work in a team?

  Being willing and able to integrate into a work group is highly valued. Show that you are willing to put personal interests aside.

- How do you rate our facility and work?

  Preliminary studies are useful for this purpose.

- What are your salary expectations?

  This is a very delicate question. Do your research beforehand, so that you can Be able to realistically assess and name your market value. Justify your desired income only on the basis of your qualifications or the benefits you can bring to the future employer. If you overprice yourself, you're out of the game. If you sell yourself short, you'll be at a financial disadvantage for a long time to come. It is often wisest to agree to the standard salary structure of the institution. In the civil service, there are clear guidelines. Only in the case of professorships and if they absolutely want you, do you have some room for negotiation, at most up to the salary of a state secretary.

And also ask yourself smart questions that announce your understanding and interest. Go to the interview rested and well-rested, so that you can also react in a relaxed manner and with presence of mind, so that your professional future is clarified positively.

But also check whether the institution you are applying to actually matches your expectations, or whether you may have got the wrong idea from afar. Because a change of job is associated with major changes and costs (relocation, new furnishings) for you and your family (often also a change of job for your spouse, a change of kindergarten or school for your children).

The above only summarises a number of essential aspects of job interviews. More detailed advice, e.g. on appearance, dress and asking questions, can be found in relevant books [41]. On the subject of application letters, which is not dealt with here, you will also find extensive literature [42].

6.3.3 Dual Career

To get a satisfying job, a scientist has to be mobile and willing to move to a new place of work, to continue his or her work at another institute, or to take on completely new challenges somewhere else. No one, especially a highly specialized scientist, can rely on finding his or her dream job at the current place of residence. Since house appointments are prohibited in Germany, taking up a professorship in particular is inevitably associated with a change of job is inevitably connected with a change of job.

But this can become a problem when one is no longer alone. If you are young and do not yet have children, it is at least also a matter of finding a job for your wife/husband or girlfriend/boyfriend – if such a bond exists – at the new place of work. If the new employer is extremely keen on your coming, there may be support in finding a second suitable job. Many universities already participate in the "Dual Career Network". But some partners may want to take a longer sabbatical and can also pass the time pleasantly abroad for one or two years as a travelling partner.

If, in addition, children have to change schools or even go abroad and the change to a foreign language takes place, the whole undertaking becomes a little more complex. But especially for the children, learning a foreign language abroad and getting used to an international climate can be a unique chance and a gain for the further life. So the decision has to be made whether the possible professional perspective is worth the effort or whether the burden on the family will not be too great. This weighing must become an additional criterion for choosing a new job when one no longer has to decide only for oneself. One's entire life should not be subordinated exclusively to one's scientific career. Scientific development should not go hand in hand with isolation.

As a compromise, many couples consider that one works alone in a remote location, for the rest of the family the usual life continues and one sees each other less often. But the reduction in immediate partner contact can become a problem for the relationship. If you see each other less than weekly at least once, perhaps only monthly or at even greater intervals, then things can soon begin to crumble in a relationship. Experience shows that about half of the relationships break up when there is constant long separation for more than two to three years. Most sufferers tell themselves this statistic doesn't have to apply to them, just like 90% of people believe they are in the top 10% of drivers who never have an accident happen to them. But you have to be aware of the risk.

Correction to: Scientific Presentation

Correction to:

Chapter 2 in: C. Ascheron, *Scientific publishing and presentation,*
https://doi.org/10.1007/978-3-662-66404-9_2

The original version of the chapter "Scientific Presentation" was inadvertently published with a restricted picture. Hence, Fig. 2.10 has been replaced with a light grey box due to legal implications.

The updated version of this chapter can be found at
https://doi.org/10.1007/978-3-662-66404-9_2

References

Ascheron, C.: Die Kunst des wissenschaftlichen Präsentierens und Publizierens: Ein Praxisleitfaden für junge Wissenschaftler. Spektrum, Heidelberg (2007) ISBN-10: 3827417414

Ascheron, C., Kickuth, A.: Make Your Mark in Science: Creativity, Presenting, Publishing, and Patents, A Guide for Young Scientists. Wiley, Hoboken (2004) ISBN-13: 978-0471657330

Lehmann, G.: Wissenschaftliche Arbeiten: Zielwirksam verfassen und präsentieren. expert, Renningen (2017) ISBN: 978-3816933755

Moser, H., Holzwarth, P.: Mit Medien arbeiten: Lernen – Präsentieren – Kommunizieren. UTB GmbH, Stuttgart (2011) ISBN: 978-3825235093, 13: 978-3825235093

Manschwetus, U.: Wissenschaftliche Arbeiten präsentieren: Leicht verständliche Anleitung für das Schreiben wissenschaftlicher Texte im Studium mit Tipps und Beispielen. Thurm Wissenschaftsverlag, Lüneburg (2016) ASIN: B01BH88CUI

Leopold-Wildburger, U., Schütze, J.: Verfassen und Vortragen: Wissenschaftliche Arbeiten und Vorträge leicht gemacht. Springer, Berlin (2002) ISBN-13: 978-3540430278

Gastel, B.: Presenting science to the public. IOP. (1983) ISBN: 978-0894950285

Todoroff, C.: Presentation Skills for Scientists, Medical Researchers, and Health Care Professionals. Trifolium Books Inc., Ontario (1997) ISBN: 978-1895579871

Alley, M.: The Craft of Scientific Presentations: Critical Steps to Succeed and Critical Errors to Avoid. Springer, Berlin (2013) ISBN: 978-1441982780

Schmale, W.: Schreib-Guide Geschichte: Schritt für Schritt wissenschaftliches Schreiben lernen. UTB GmbH, Stuttgart (2006) ISBN-13: 978-3825228545

C. Ascheron, *Scientific publishing and presentation*,
https://doi.org/10.1007/978-3-662-66404-9

Mylonas, I., Brüning, A.: Wissenschaftliches Publizieren in der Medizin: Ein Leitfaden 1st Edition, Kindle Edition. Springer, Berlin (2013) ASIN: B00GXJXTNC

Erdnüß, F.: Wissenschaftliche Paper publizieren für Dummies. Wiley-VCH, Weinheim (2016) ASIN: B01E0JTYUI

Chirlek, G., Wanner, I.: Wissenschaftliches Schreiben und Publizieren: Erläuterung für Studierende und Doktoranden. Books on Demand, Norderstedt (2015) ASIN: B00EPR6IM8

Ebel, H.F., Bliefert, C., Greulich, W.: Schreiben und Publizieren in den Naturwissenschaften. Wiley-VCH, Norderstedt (2012) ASIN: B0096QYQUY

Weingart, P., Taubert, N.: Wissenschaftliches Publizieren: Zwischen Digitalisierung, Leistungsmessung, Ökonomisierung und medialer Beobachtung. De Gruyter, Berlin (2016) ISBN: 978-3110448108

Fischer, S.: Erfolgreiches wissenschaftliches Schreiben. Kohlhammer, Stuttgart (2014) ASIN: B00QMH5KXM

Oehlrich, M.: Wissenschaftliches Arbeiten und Schreiben: Schritt für Schritt zur Bachelor- und Master-Thesis in den Wirtschaftswissenschaften. Springer Gabler, Wiesbaden (2014) ASIN: B00UZBE2V2

Kornmeier, M.: Wissenschaftlich schreiben leicht gemacht: Für Bachelor, Master und Dissertation, 7th edn. UTB GmbH, Stuttgart (2016) ASIN: B01EV71IR6

Schmölzer-Eibinger, S., Bushati, B., Ebner, C.: Wissenschaftliches Schreiben lehren und lernen: Diagnose und Förderung wissenschaftlicher Textkompetenz in Schule und Universität. Waxmann, Münster (2018) ISBN-13: 978-3830937692

Huss, J.: Schreiben und Präsentieren in den angewandten Naturwissenschaften: Leitfaden für die Anfertigung von Diplomarbeiten und Dissertationen in der Forstwissenschaft und verwandten Fachgebieten. Norbert- Kessel-Verlag, Remagen (2014) ISBN: 978-3941300941

Manschwetus, U.: Quellen richtig zitieren: Leicht verständliche Anleitung für das Schreiben wissenschaftlicher Texte im Studium mit Tipps und Beispielen. Thurm Wissenschaftsverlag, Lüneburg (2016) ASIN: B019B9NW8W

Hayden, T., Nijhuis, M.: The Science Writers' Handbook: Everything you Need to Know to Pitch, Publish, and Prosper in the Digital Age. Da Capo Lifelong Books, Boston (2013) ASIN: B06XCP5BJW

Blackwell, J., Martin, J.: A Scientific Approach to Scientific Writing. Springer, Berlin (2011) ASIN: B008CCGPA2

Körner, A.M.: Guide to Publishing a Scientific Paper. Routledge, London (2008) ASIN: B001QEQR6U

Huber, K.-P.: Die Doktorarbeit: Vom Start zum Ziel: Lei(d)tfaden für Promotionswillige, Barbara Messing. Springer, Berlin (2007) ASIN: B00BH8469K

Knigge-Illner, H.: Der Weg zum Doktortitel: Strategien für die erfolgreiche Promotion. Campus, Frankfurt (2015) ASIN: B0113ABJG4

Hell, S.: Soll ich promovieren? Voraussetzungen, Chancen und Strategien. Vahlen, München (2017) ASIN: B07897Y5T8

Hackstein, H.: Doktorarbeiten in Medizin und Life Sciences: Schnell und gut verfassen: Tipps und Tricks – auch für Masterstudenten. Amazon Media EU. ASIN: B00VD3JT1Y

Anja, S.: So promovieren Sie richtig: Der Leitfaden zum Doktortitel. Amazon Media EU. ASIN: B00Z1XMD2E

Wolf, H.: Dissertation in 30 Tagen: Das Praxisbuch für die medizinische Promotion. Amazon Media EU. (2017) ASIN: B076JMVK5L

Feibelman, P.J.: PhD Is Not Enough! A Guide to Survival in Science. Basic Books, New York (2011) ASIN: B06XCG2ZQQ

Allison, B., Race, P.: The Student's Guide to Preparing Dissertations and Theses. Routledge, London (2004) ASIN: B000P0JMZY

Gosling, P., Noordam, L.D.: Mastering Your PhD: Survival and Success in the Doctoral Years and Beyond. Springer, Berlin (2010) ASIN: B008BFCY32

Woodhouse, I.: 101 Top Tips for PhD Students. Speckled Press, Berlin (2015) ASIN: B00YB9BTXI

Osawa, E.: Kagaku. **25**, 843 (1970)

von Klitzing, K.: Phys Rev Lett. **45**(6), 494–497 (1980)

Bednorz, J.G., Müller, K.A.: Z. Phys. **64**, 189 (1986)

APS Guidelines for Professional Conduct. https://www.aps.org/policy/statements/02_2.cfm

Kohler, M.: Entspanntes Atmen bei Yoga, Albert, Müller, Stuttgart (1972). ASIN: B001U1C0ME

Iding, D.: Die heilende Kraft des bewussten Atmens: Vital, ausgeglichen und entspannt im Alltag. Knaur TB, München (2004) ISBN-13: 978-3426872215

Püttjer, C., Schnierda, U.: So überzeugen Sie im Bewerbungsgespräch – Die optimale Vorbereitung für Hochschulabsolventen. Campus, Frankfurt (2005)

Neumayer, G., Rudolph, U.: Das Bewerbungsschreiben. Humboldt, Hannover (1999)